THE DRUG HUNTERS

The Improbable Quest to Discover New Medicines

DONALD R. KIRSCH
AND OGI OGAS

FOREWORD BY MADELYN FERNSTROM

PREFACE TO THE PAPERBACK BY DONALD R. KIRSCH

Arcade Publishing • New York

First Paperback Edition 2018

Arcade Publishing books may be purchased in bulk at special discounts for sales promotion, corporate gifts, fund-raising, or educational purposes. Special editions can also be created to specifications. For details, contact the Special Sales Department, Arcade Publishing, 307 West 36th Street, 11th Floor, New York, NY 10018 or arcade@skyhorsepublishing.com.

Arcade Publishing® is a registered trademark of Skyhorse Publishing, Inc.®, a Delaware corporation.

Visit our website at www.arcadepub.com.
Visit the authors' site at www.TheDrugHunters.com.

10 9 8 7 6 5 4

Library of Congress Cataloging-in-Publication Data

Names: Kirsch, Donald R., 1950– author. | Ogas, Ogi, author.
Title: The drug hunters : the improbable quest to discover new medicines / Donald R. Kirsch, PhD and Ogi Ogas, PhD.
Description: First edition. | New York : Arcade Publishing, [2017]
Identifiers: LCCN 2016034624 | ISBN 9781628727180 (hardback) | ISBN 9781628729863 (paperback) | ISBN 9781628727197 (ebook)
Subjects: LCSH: Drugs—Research—History. | Pharmacology—History. | BISAC: SCIENCE / History. | MEDICAL / Pharmacology. | MEDICAL / History. | BIOGRAPHY & AUTOBIOGRAPHY / Medical.
Classification: LCC RM301.25 .K57 2017 | DDC 615.1072—dc23 LC record available at https://lccn.loc.gov/2016034624

Cover Design: Erin Seaward-Hiatt
Cover photo: Slobodan Cedic

Printed in the United States of America

Contents

FOREWORD

Madelyn Hirsch Fernstrom, PhD

We all take for granted the process of filling a prescription. The right medicine targets our specific ailment and we're confident that we will recover quickly. But none of us really thinks about where this medicine comes from. *The Drug Hunters* provides quite an interesting and insightful read in answering this question. We are led through a complex journey of several hundred years of drug discovery—from casual (sometimes brilliant) observation, to just plain luck, and ultimately to intelligent drug design. We share the experiences of Drs. Kirsch and Ogas, as they provide a bird's eye view of the creation of the field of pharmacology—discovery, standardizing, and cataloging medicines for both safety and efficacy.

It's hard to believe that until the discoveries of Louis Pasteur and the development of the "germ theory" around 1860, there was not even a clue that microorganisms (such as bacteria) caused illness. And it took an additional forty years before the medical community came to accept this idea. Doctors simply didn't believe, for example, that hand washing could prevent disease. When they finally did incorporate this simple action, patient infection rates dropped precipitously.

Although the medical community was often resistant to change, usually for fear of harming a patient, drug pioneers forged on to document the benefits of new thinking about the developing field of "medications," and society has been dramatically changed by the development of those drugs. After all, it was only two generations ago that patients with tuberculosis were separated and housed in sanitariums as the only form of "treatment." Without effective medications, psychiatric patients were also placed in long-term care hospitals to prevent them from hurting themselves and others. Millions of people died from the consequences of high blood pressure.

Looking at treatments used in the past, it's easy to understand just how lucky we are to have the modern medications we rely on. Before the discovery of penicillin—the first effective antibiotic—millions of people died from bacterial infections that started small, and then spread throughout the body. Infections that today are routinely treated with antibiotics like strep throat, rheumatic fever, and some pneumonias, were deadly before then. This discovery which changed the course of history wasn't the result of months of targeted research for an antibiotic, but rather the accidental observation by Alexander Fleming that his bacterial samples were infested with a fungus that killed them. The process continued for another twenty years as scientists relied on skill and even more luck to develop the widely available medicine we rely on today. Even this victory was short-lived, though, as luck isn't always on the drug hunters' side. With continued study scientists realized that the bacteria were changing, becoming resistant to the drugs, and so the quest for new antibiotics continued.

The Drug Hunters clearly demonstrates the idea that there is no rational process for drug discovery. There is no clear set of "scientific laws, engineering principles, or math formulae that can guide an aspiring drug hunter . . . from idea to product." Even the modern era of drug discovery, which exploded in the 1950s when large numbers of new agents miraculously appeared, development was generated in large part by accident and aided by vast new scientific knowledge about "what can go wrong" in the body (pathophysiology). The continuing stream of new discoveries about how the body works gives us fresh ideas about where to look for bioactive molecules and their mechanisms of action. This ongoing accumulation of knowledge suggests ways to target new drugs and develop them. The skill of a drug hunter is in their dedication to balancing this ever-expanding knowledge and their tenacity in exploring new possibilities, all while hoping for a moment of serendipity.

If you've ever taken a medication—or plan to in the future—read this book. You will gain a new appreciation for where your medications come from and why they work!

Preface

Most people believe medicines are high-technology inventions discovered by eccentric scientists toiling away in super-advanced, well-equipped laboratories. It is true that first-rate science is necessary for drug discovery, but pure technical prowess is not sufficient. Teamwork and collaboration are just as, if not more important than sophisticated science for the discovery of new drugs. The most recent drug to have been discovered by a single individual was the surgical anesthetic ether, discovered in 1846 by William T. G. Morton. Since then, all new drugs have been discovered by teams.

The importance of teamwork and collaboration is unfortunately often neglected in scientific training, and thus drug hunters commonly must learn how to collaborate on the job. I was lucky to have many people at various stages of my career help me become a better team member and collaborator, and I'm so very thankful for their efforts and treasure the relationships I developed with them.

Richard Sykes, my first boss in the pharmaceutical industry, was a strong and charismatic leader who had a great influence on my career. Leaders are often thought of as powerful people who direct others and make important decisions. But the best leaders are powerful mainly because they empower those who follow their lead: they become so by sharing their power. My first boss delegated work and decisions to his team to make good things happen. His confidence in me gave me confidence in myself, made me feel like I could accomplish important things. The confidence he instilled was an important force throughout my career. It enabled me to try new approaches, take considered risks, and tackle difficult tasks. Among many other things, it gave me the courage to write this book.

Early in my career, I was a poor team member, and was often thoughtless regarding the needs of others on the team. Fortunately, a senior scientist colleague at Squibb named Phil Principe generously took me under his wing and mentored me on how to work productively with others. He helped me to see projects from viewpoints other than my own.

After group meetings, he would often take me aside and say, "Donald, you have to work on your interpersonal relationships." At first, I thought he was nuts. I was trying to get the team to pursue the very best science. What was wrong with that? But I later came to realize that forcing your own ideas down people's throats was not the best way to get people to work with you. I don't think I could have become a better team member without Phil's help.

A personal tragedy in Phil's life led him to become highly compassionate toward others. His wife died from bladder cancer when she was a young woman, leaving him with three small children to raise alone. During the dark days after his wife's death, he would come home from work, pour himself a glass of whisky, and hug his children until he composed himself. "Kids, kids, kids," he would say over and over again. His kids saved him—plus two fingers of

Scotch. He showed considerable compassion toward me, sharing his more than thirty years of drug discovery experience when I was just a young, green-as-grass novice.

In the mid-1990s my lab in New Jersey started collaborating with a research subsidiary group in Germany. I felt that the head of the German biology group had no faith in the approaches we were taking. But he was a great team player. He was open-minded and, although his own preferred approach was totally different from mine, for the good of the collaboration he tried his very best to see the positives in what my group was doing. Despite his own views, he would consider our scientific strategies and approaches without letting his own opinions get in the way.

He was also a kind host whenever our team visited the German lab and would show us the local sights. He even spent time trying to teach me German. Under his tutelage, my German improved from totally incomprehensible to only mostly incomprehensible: a significant feat.

I later would travel to Japan on business. My company had an office in Tokyo, and our Japanese representative could not have been more giving or a better colleague. Ando-san could have carried out his job effectively by simply working with me during normal business hours. This would have been totally reasonable, especially considering that he had a two-hour commute each way from his suburban apartment to our offices in the Roppongi district of Tokyo, while I merely needed to walk a few blocks to the office from a nearby hotel. Yet he was always at his desk when I arrived in the morning and never seemed to be anxious to return home at the end of the day. He always made sure that everything was taken care of for me before he left for home. And in the evenings, he would generously show me interesting sections of Tokyo and take me to sample all sorts of exotic Japanese cuisine.

One of the greatest contemporary drug discovery scientists and a personal hero of mine is Satoshi Ōmura. Dr. Ōmura's crowning achievement was his work on ivermectin, a drug that has cured millions of people suffering from disfiguring and life-threatening parasitic diseases: river blindness, lymphatic filariasis, and Guinea worm disease. In 2015, Dr. Ōmura was awarded the Nobel Prize in Physiology or Medicine for his role in the discovery of this important medicine.

In 1986 Dr. Ōmura wrote a paper about his drug discovery philosophy:

"I would like to present my viewpoint and ideas on research work. These achievements have been produced by (i) believing in the great capabilities of microorganisms, (ii) establishing a well-designed screening system for desired substances, (iii) recognizing that screening is not just routine work, (iv) emphasizing basic studies, and (v) treasuring good human relationships."

These ideas, appreciating the wonder and power of nature, focusing research efforts on revealing the secrets that the natural world holds, never losing your sense of wonder, and finally, and most importantly, striving to become the best team member you can be, have very much guided my own career.

I was fortunate to have had good colleagues throughout my career, and hopefully they taught me to be a good colleague as well. Whatever you do, it is important to remember the words of Dr. Ōmura: treasure good human relationships.

DONALD R. KIRSCH

THE DRUG HUNTERS

Introduction
Searching the Library of Babel

The Library of Babel

"By this art you may contemplate the variations of the 23 letters. . . ."

—Jorge Luis Borges, "The Library of Babel"

In the deep mists of prehistory, everybody was a drug hunter. Our parasite-infested, malady-ridden ancestors chewed every new root and leaf they chanced upon, hoping to unleash some fortuitous benefit that might ameliorate their afflictions—and praying they did not perish from their blind experimentation. Through sheer serendipity, some fortunate Neolithic souls stumbled upon substances with medicinal properties, including opium, alcohol, snakeroot, juniper, frankincense, cumin, and—apparently—birch fungus.

Sometime around the year 3300 BC, a solitary man, cold, ill, and mortally wounded, struggled through the high peaks of the Ötztal Alps in Italy until he collapsed in a crevasse. Here he lay frozen for more than five thousand years until hikers stumbled upon his ice-mummified corpse in 1991. They dubbed him Ötzi. When Austrian scientists thawed out this Ice Age hunter, they discovered his intestines had been infested with whipworms. At first the researchers remarked that Ötzi and his contemporaries most likely suffered the indignities of this painful parasite without any hope for relief. A second discovery prompted them to revise their convictions.

Attached to Ötzi's bearskin leggings were two hide strips, each knotted around a rubbery white lump. These strange bulbs turned out to be fruiting bodies of the birch polypore, a fungus with antibiotic and anti-hemorrhaging properties. It also contains oils that are toxic to whipworms. Ötzi's hide-knotted mushrooms are quite likely the oldest medicine kit ever found. The Iceman's medicines did not have high potency or efficacy—but they worked. The existence of a five-thousand-year-old anti-worm drug (what pharmacologists would call an antihelminthic) reminds me of something my PhD advisor used to say: "When you see a dog walking on its hind legs, you are not impressed by his grace or agility but rather by the fact he can do it at all."

Ötzi's remarkable birch fungus embodies a simple truth about humankind's quest for medicine. This Neolithic remedy did not arise from clever innovation or rational inquiry. No Stone Age Steve Jobs engineered this antihelminthic out of the visionary workings of his mind. No, Ötzi's drug was the product of sheer unadulterated luck. All prescientific drug hunting advanced through simple trial and error.

And today? As Pfizer, Novartis, Merck, and other Big Pharma conglomerates spend billions of dollars on state-of-the-art drug hunting laboratories, you might guess that most blockbuster drugs are the fruits of meticulously planned drug engineering projects where the role of trial-and-error has been replaced with informed scientific execution. Not so. Despite the best efforts of Big Pharma, the prime technique of the twenty-first-century quest for medicine remains the same as it was five millennia past: painstakingly sampling a mindboggling variety of compounds and hoping that one of them, *just one*, proves out.

Over the course of my nearly four-decade career as a drug hunter, I've learned firsthand that new medicines are often discovered by routes that are wildly circuitous or entirely fortuitous—or both. The professional drug hunter is kin to the professional poker player: possessing enough knowledge and skill to tilt the game in his favor at crucial moments but always at the mercy of the shuffle of the cards.

Consider rapamycin. In the 1970s, biologist Suren Sehgal was working for Ayerst Pharmaceuticals looking for a new drug to treat common fungal infections such as Candida vaginitis and athlete's foot. After sampling tens of thousands of compounds, Sehgal discovered a novel antifungal compound that originated in a soil microorganism found on Easter Island. He named the drug rapamycin after Rapa Nui, the aboriginal name for the remote Pacific island.

Sehgal tested rapamycin on animals and found that it wiped out any malevolent fungi. Unfortunately, the drug also suppressed the animals' immune system. If you are trying to eliminate an infection, especially a fungal infection, it is crucial for the immune system to work effectively and in concert with an antifungal medicine. This unfortunate side effect proved to be insurmountable, and the Ayerst executives decided to ditch rapamycin and move on.

But Sehgal did not want to give up. He knew that another antifungal compound, cyclosporine, was being developed for a very different use—as an organ transplant therapy. Like the Easter Island drug, cyclosporine also produced immunosuppressive effects, but this was a desirable property for a post-transplant drug because it helps prevent the body from rejecting the new organ. Sehgal reasoned that rapamycin might also be useful as an anti-rejection therapy.

Unfortunately, his employer (which by that time had merged with another company—a frustratingly common occurrence in my industry) did not possess an immune suppression research program, and since the new management team was not interested in organ transplants they dismissed Sehgal's proposal out of hand. But Sehgal, a seasoned drug hunter, was well aware of one of the most reliable facts about Big Pharma: rapid executive turnover. He bided his time. Whenever a new management team assumed control over the pharmaceutical research, he reintroduced his proposal to test rapamycin as an organ transplant therapy.

On the third or fourth such occasion, Sehgal's boss became annoyed by what he perceived as Sehgal's incessant nagging in the pursuit of a futile pet project. His boss ordered him to take his culture of Easter Island bacterium, dump it in an autoclave, and hit the sterilize button. This would destroy the microorganism once and for all, along with Sehgal's dreams of a transplant drug—or at least, that's what his boss hoped. Sehgal did comply with his manager's demands . . . but only after taking a rapamycin culture home with him and storing it in his freezer, perhaps squeezing it between his veal cutlets and frozen peas.

Sehgal's gamble paid off. Exactly as he hoped, his boss soon moved on to another job and yet another management team took the reins. And once again, Sehgal pitched rapamycin as an anti-rejection drug. This time, his pitch worked. The new execs gave his long-mothballed project the green light. Sehgal yanked the culture back out of his kitchen freezer, re-created the drug, then tested it on transplants in animals . . . success! Finally, he tested it on actual transplant patients . . . victory! In 1999—about twenty-five years after Sehgal first discovered it—the Easter Island anti-fungal drug was finally approved as an immunosuppressive agent by the FDA. Today it is one of the most commonly used anti-rejection therapies. It is also used as a coating for coronary artery stents to increase their longevity, a surprising outcome for a drug that was originally intended to fight athlete's foot and yeast infections.

Or perhaps it is not so surprising at all. After spending my entire career searching for new medicines, I've learned that the only sure thing in the drug hunting business is that you almost never end up with the exact drug you started stalking. The vast majority of my colleagues, all educated at top-flight research universities and working at posh laboratories festooned with high-tech gear, have spent their entire careers groping through the labyrinth of bio-active molecules without ever finding a new compound that safely and effectively improves human health.

The professor who trained me in pharmacology, an MD, once told me that 95 percent of the time a patient visits a physician he will not actually be helped by the doctor. In most cases, either the patient's body will heal itself without needing the doctor's intervention or the disease will be untreatable, rendering the physician powerless. In his view, the physician has the ability to make a meaningful difference to his patients only 5 percent of the time. While that may seem low, those are fantastic odds compared to the ones faced by the drug hunter.

Only about 5 percent of a scientist's ideas for a drug discovery project get funded by management. Of these, only 2 percent end up producing an FDA-approved medicine. That means a drug hunting scientist can only expect to make a difference about one-tenth of 1 percent of the time. Finding new drugs is so challenging, in fact, that it has led to a crisis in the pharmaceutical industry. Big Pharma companies are becoming increasingly frustrated with the massive research expenditures necessary to come up with new drugs—an average of about $1.5 billion and fourteen years for each FDA-approved drug—and the exasperating fact that the vast majority of their endeavors don't produce a usable drug. Executives at Pfizer recently told me they are thinking about getting out of the drug-discovery industry entirely. Instead, they want to be in the drug *acquisition* industry: they would prefer just to buy drugs that other people have invented. Think about that. Finding new drugs is so formidable that one of the oldest, most talented, and wealthiest drug makers—the largest drug maker in the world, in fact—would rather let other people deal with the problem.

So why is the "degree of difficulty" of finding new drugs so much greater than, say, landing a man on the moon or designing an atomic bomb? The moon shot and the Manhattan Project employed well-established scientific equations, engineering principles, and mathematical formulas. They were formidable and grueling endeavors, to be sure, but at least the researchers possessed clear scientific road maps and mathematical compasses to guide them. The moon shot engineers knew with certainty the distance from the Earth to the moon, and how much fuel was needed to get there. The Manhattan Project scientists knew that matter could be turned into city-obliterating energy according to $E=mc^2$.

In contrast, the core challenge of designing a new drug—the trial-and-error screening of immense numbers of candidate compounds—is a task not guided by any known equations or formulas. While an engineer knows if his bridge will bear weight before he ever lays a girder down, a drug hunter has no clear idea how a particular drug will work until a human subject actually ingests it.

In the mid-nineties, chemists at Ciba-Geigy (now a part of Novartis) calculated the total number of possible drug compounds in our universe: 3×10^{62}. When it comes to characterizing the size of a number, some numbers are big, some are enormous, and some are so incomprehensibly, inconceivably large that they might as well be infinity. The number 3×10^{62} falls into that third category. If you were able to test one thousand compounds every second to see if any of them could serve as an effective remedy for a particular malady—say, breast cancer—by the time our sun burned out you would still have not made a measurable dent in the total number of breast cancer-fighting drug possibilities.

There is a story by the blind Argentinean author Jorge Luis Borges that I think perfectly captures the central challenge of drug hunting. In "The Library of Babel," Borges imagines that the universe is a library consisting of an infinite number of hexagonal rooms that extend forever in every direction. Each room is filled with books. Each book contains a random arrangement of letters, and no two books are the same. Once in a while, purely by chance, a book contains an entire readable sentence, such as "The gold is in the mountain." But, as Borges puts it, "For every rational line or forthright statement there are leagues of senseless cacophony, verbal nonsense, and incoherency."

Nevertheless, the library *must* contain books that, purely by chance, are filled with legible life-changing wisdom. Such books are known as "Vindications." In Borges's fantasy, solitary searchers known as librarians wander endlessly through the library, hoping to find these Vindications. Most librarians wander through the endless hexagons in vain, spending their life coming across nothing but nonsense. But Borges notes that there are librarians who, through good fortune or sustained force of will, have managed to discover a Vindication.

Similarly, every possible drug is contained somewhere in the vast theoretical library of chemical compounds. There is a molecular configuration that will safely destroy ovarian cancer, another that will halt the corrosive advance of Alzheimer's, another that will cure AIDS—or maybe they do not exist at all. There is no way to know for sure. Modern drug hunters are like Borges's Babelian librarians, forever questing for life-changing compounds and always suppressing their secret fear that the vindicating medicines may never be found.

The problem, ultimately, is the human body. Our physiological activity is not a closed, well-defined system like rocket propulsion or nuclear fission. It is an open and unfathomably arcane molecular system with innumerable undefined relationships among its components, rendered even more abstruse by the fact that each person's body has their own idiosyncratic structure and dynamics. We only understand a tiny fraction of these physiological relationships and have not yet deciphered how most of our body's basic molecular components actually work. Complicating matters still further is the fact that each individual has her own idiosyncratic genetic and physiological architecture, so that each person's body operates slightly (or extremely) differently. Even more daunting, despite tremendous advances in our understanding of cells and tissues and organs, we simply cannot precisely predict in advance how a given chemical compound will interact with a given molecule in a living body. In fact, it is impossible to know with certainty if a particular disease possesses what pharmacologists call a "druggable protein" or a "druggable target"—some specific protein associated with a pathology that can be influenced by a chemical agent.

Designing an effective drug requires two things: the right compound (the drug) and the right target (the druggable protein). The drug is like a key that turns the protein lock to start the ignition on a physiological engine. If a scientist wants to intentionally influence a person's health in a specific way—to reduce depression, relieve itching, treat food poisoning, or produce any other health benefit—she must first identify a target protein that influences the relevant physiological processes in the human body or that, conversely, interferes with the physiological processes of a pathogen.

For example, Lipitor acts upon HMG-CoA reductase, the protein controlling the rate of the body's synthesis of cholesterol. Penicillin, in contrast, shuts down peptidoglycan transpeptidase, a protein required to synthesize the (essential) cell wall of a bacterium. But finding the drug key that will turn a protein lock . . . As Hamlet would say: *Ay, There's the rub!* This is the daunting challenge for the drug hunter. Despite the humbling odds, some drug hunters, such as Suren Sehgal, through unwavering resolution or outrageous fortune, through individual genius or far-flung collaborations, have stumbled upon their Vindications.

The term drug hunters have bestowed upon the process of systematically searching through a library of compounds is *screening*. The prehistoric screening method consisted of snatching every new berry or leaf you came across and snorting it, smearing it, or swallowing it. After unknowable eons of our ancestors randomly sampling the natural landscape, in 1847 the first drug was discovered using a reasonably scientific method of screening. At the time, physicians were using ether as a surgical anesthetic, prompting them to reason that there could be other chemical compounds similar to ether that might work even better. Ether had a few obvious shortcomings—it irritated patients' lungs and had an unfortunate tendency to explode—so physicians knew there would great clinical value for a new anesthetic without these issues.

Since ether was a volatile organic liquid, the Scottish physician James Young Simpson and two of his colleagues decided to test every volatile organic liquid they could get their hands on. Their screening process was simple: they opened a bottle of a given test liquid and inhaled its vapors. If nothing happened, they labeled the sample *inactive*. If they woke up on the floor, they labeled the sample *active*.

This screening protocol, of course, would never meet contemporary standards for laboratory safety. Benzene, for instance, is a volatile organic liquid that was widely available at the time and was almost certainly one of the compounds that Simpson screened. We now know that benzene is carcinogenic, and inhaling its vapors can cause long-term damage to your ovaries or testes.

Despite the recklessness of their screening method, on the evening of November 4, 1847, Dr. Simpson and his colleagues tested chloroform. When the three men inhaled the chemical it produced a mood of cheer and good humor—followed by collapse and unconsciousness. When they awoke hours later, Simpson knew they had identified an active sample.

Hoping to confirm their findings, Simpson insisted that his niece, Miss Petrie, inhale the chloroform while he watched. The girl blacked out. It is fortunate she woke again, since we now know that chloroform is a powerful cardiovascular depressant, producing a high incidence of death when used as a surgical anesthetic. Despite these dangers, by sniffing chemical after chemical in his living room, Simpson had discovered one of the blockbuster drugs of the nineteenth century—a pharmaceutical origin story unlikely to be duplicated today. But you never know. In the 1980s, I tried to find new drugs in the back of a Volkswagen microbus.

If you are thinking that I must have been indulging in tie-dyed psychedelic experimentation—after all, why else would anyone be diverting himself with unknown drugs in the back of a lime-green VW bus?—you would be wrong. One of my first paid jobs was working as a drug hunter in an antibiotics discovery group. A common method for searching for new antibiotics is to screen microorganisms living in the soil. I always kept an eye out for new kinds of soil that might hide a pharmaceutical payoff—and a commercial payoff. I was literally looking for pay dirt.

One weekend, I volunteered to travel to the Delmarva Peninsula to screen soil samples from the Chesapeake Bay side of the peninsula. I took our "mobile laboratory"—the microbus, which we had equipped with a sink and a Bunsen burner. Since my group had recently discovered a new type of antibiotic called monobactams, we christened our mobile laboratory the "Monobacvan."

I somehow managed to recruit my wife to come with me with promises of sunbathing on the beach, but then conscripted her into driving the Monobacvan around the tight curves of the rural shoreline as I hunkered down in the back, abruptly ordering her to stop so I could dash out and fill bags full of dirt. During those moments when we were not driving or scooping up the dank, smelly Chesapeake earth, I was diluting the samples and slapping them on Petri plates. My wife was not pleased. The weekend was a bust for both of us, since when I got back to the lab on Monday and we tested my samples, each one turned out to be inactive. My wife informed me that if I did not want my marriage labeled inactive, our next road trip needed much more sunbathing and absolutely no more screening.

When people learn that I am a drug hunter they usually ask me at least one of the following three questions—usually expressed with some well-founded cynicism:

Why are my drugs so expensive?

Why do my drugs have such unpleasant side effects?

Why is there no medicine for the malady that afflicts me or someone I love?

One reason I wrote this book was to answer these questions, and the truth of the matter is that the answers to all three are tied to the fact that drug hunting—at least, thus far—is dismayingly difficult because every contemporary method of drug development relies, at some crucial juncture, on trial-and-error screening, just as it did when Neanderthals roamed the wilds. We still do not possess adequate knowledge of human biology to provide us with theories and principles that could rationally guide us to the salubrious molecules we so fiercely desire.

But as I started working on the book, I realized there were even deeper lessons to share about human health, the limits of science, and the value of courage, creativity, and inspired risk-taking. In the chapters that follow, I will share the sweeping journey our species has undertaken in the quest for medicine from our Stone Age forebears to today's Big Pharma megacorporations, chronicling humankind's search for elusive cures concealed somewhere within the near-infinite library of chemistry. I've tried to write in an accessible fashion that non-scientists can easily follow, putting more technical observations in the end notes—along with interesting details and anecdotes that didn't quite fit into the book's general flow. I will narrate this epic adventure by telling the stories of those remarkable individuals whose intuition, innovation, perseverance, and remarkable good fortune led them to their Vindication. On the way we will try to discern what lessons they may hold for the future of our well-being. What enabled history's most successful drug hunters to discover world-changing medicines? And is there anything we can do, individually or as a society, to boost the odds of findings the drugs we need the most?

In addition to these lofty goals, I confess I also nurse a more personal and modest mission for this book, the original inspiration that first motivated me to sit down and write. I want to share with you, in unvarnished fashion, what it's like to be a professional drug hunter.

1 So Easy a Caveman Can Do It

The Unlikely Origins of Drug Hunting

A field of poppies

"Among the remedies which it has pleased Almighty God to give to man to relieve his sufferings, none is so universal and so efficacious as opium."

—Thomas Sydenham, seventeenth-century English physician

Our prehistoric forebears harbored an extravagant menagerie of supernatural beliefs. They believed it was possible to prepare potions from flowers that would hide you from the spears of your enemies. They thought snorting pulverized twigs would bestow the power to hear your neighbor's thoughts. They also believed, with equal improbability, that foul-smelling concoctions brewed from tangly roots would cure disease.

Today, we find the suggestion that a chemical might endow you with invisibility or telepathy to be preposterous. On the other hand, we do not bat an eyelash at the prospect of finding healing salves within the natural world—in fact, we take Mother Nature's bountiful pharmacopeia for granted. Yet why should the notion of botanical cures be any less outrageous than the notion of botanical telepathy? If you think about it for a moment, why in the world would the juice of a pungent bark found in mucky swamplands hold the power to relieve arthritis or promote digestion or lower blood pressure in *Homo sapiens*?

Certainly if one imagines that all the world was created for the express benefit of humankind, with all the flora and fauna of the Earth designed by a munificent deity to nourish our divinely ordained species, then we might believe it was God's will for the sap of the willow tree to assuage headaches and the leaves of the foxglove plant to relieve heart ailments. However, if we believe in the principles of evolutionary biology, then it is far more surprising—mystifying, even—that so many compounds produced in non-human species have a salubrious effect on our own kind.

We cannot be certain that the impulse that drove the earliest humans to rifle through the leafy shelves of Mother Nature for new medicines was the same one that drove them to seek out invincibility powders and clairvoyance potions, but we do know that even the most primitive of human beings somehow plucked out effective drugs, like Ötzi's parasite-slaying fungus.

It is not too hard to imagine that a plant substance might kill a parasite or even a bacterium; after all, many creatures produce toxins to defend themselves against infestation. But what about a plant that assuages our pain or ameliorates our acne? Or, even more peculiar, a plant that improves our mood or expands our consciousness? It is difficult for our modern minds, used to the overflowing aisles of candy-colored pills and syrups at the local Walgreens, to appreciate just how unlikely and bizarre organic drugs truly are—but what if I told you that there was a bush whose berries, when eaten, would enable you to breathe underwater? There's not, of course, but we should feel equal skepticism and astonishment regarding the fact that the plant kingdom produces compounds that benefit our animal bodies in ways that have absolutely nothing to do with the way the compounds operate in plants.

Somehow, natural remedies were exhumed and harnessed by our prehistoric progenitors, even as their understanding was laced with myth and magic. Remarkably, some of these Stone Age drugs have stood the test of time and remain in widespread use today. Opium is one. Tracing the history of opium, one of humankind's most ancient medications, will illustrate just how perplexing the existence of naturally occurring drugs really is, while also serving as a useful introduction to our species's venerable quest for medicine.

If we relegate alcohol to the status of a beverage, then the oldest known medicine is something that you and I and almost every person in Western civilization has imbibed at some point in our lives—the tincture of the poppy. Percocet, morphine, codeine, oxycodone, and (of course) heroin are all derived from *Papaver somniferum*, a wild plant with attractively colored flowers common to Asia Minor. Opium is the active ingredient of the poppy, and one reason the drug has been used for so long is because opium is incredibly easy to prepare: the immature fruit of the poppy plant is scratched so that sap oozes out, the sap is collected, dried, and ground into a powder—and voilà, pure opium.

Opium was used by the Sumerians as early as 3400 BC, who referred to it as *Hul Gil*, the "joy plant." The Sumerians passed along their knowledge of the joyful effects of the poppy to the Assyrians, who passed it on to the Babylonians, who passed it to the Egyptians. The first-known reference to poppy juice appears in the writings of the Greek philosopher Theophrastus in the third century BC; the word opium comes from the ancient Greek word for "juice," *opion*. Later, Arabian traders introduced opium to Asia, where it was used in the treatment of dysentery, an often-fatal disease characterized by explosive diarrhea; in addition to its narcotic effects, opium is also highly constipating.

One major limitation of opiates as a medication is their poor solubility in water. After four thousand years of repeating the same simple water-based preparation of opium, many physicians in the

Middle Ages attempted to develop a more effective formulation. These men represented one of the earliest breeds of drug hunters, the "formulator"—someone who tried to figure out a new way to prepare a known medicine. These formulators relied upon crude knowledge of prescientific chemistry, the pseudoscience of alchemy, or indiscriminate experimentation to develop new mixtures that often contained as many non-active compounds as active agents.

Paracelsus, a sixteenth-century botanist–physician, was one of the most talented of the formulator drug hunters. He devised a novel preparation of opium by dissolving it in alcohol. The preparation came to be known as laudanum, though Paracelsus himself was so enamored of its powers that he referred to it as the "stone of immortality." His alcohol-based version of opium came close to pharmaceutical immortality, as it was still being used late into the twentieth century.

Another alcohol-based opiate mixture is known as paregoric. First formulated in the eighteenth century by Le Mort, a chemistry professor at the University of Leiden, paregoric is familiar to readers of Victorian novels because their heroines were frequently dosed with paregoric to calm frazzled nerves after some social drama, such as rejection by the handsome young baron. The word paregoric, in fact, comes from the Greek term for "soothing."

Dover's Powder, another eighteenth-century opium preparation, was invented by Thomas Dover in 1732. While scientists know Thomas Dover as an early pharmacologist, he became famous to the public through his other adventures. After studying medicine at Cambridge University, Dover settled in the English port city of Bristol and at the age of fifty joined a privateering venture to the South Seas. In 1709, the expedition landed on a deserted island off the coast of Chile—except Dover and his party discovered that the island was not deserted after all. It was inhabited by Alexander Selkirk, the sole survivor of a shipwreck from four years earlier. Upon Selkirk's return to England, he became a celebrity, and his dramatic story inspired Daniel Defoe to write *Robinson Crusoe*. Upon Dover's own return to England, however, he invented Dover's Powder, coarse off-white granules containing equal quantities of opium and ipecac, an ingredient once found in cough syrup. His newly acquired fame as Selkirk's rescuer surely did him no harm in marketing his new remedy.

Opium itself is actually a complex mixture of many different active compounds, such as the phenanthrenes (which includes familiar analgesics like morphine and codeine) and the benzylisoquinolines (which includes papaverine, a drug used as a treatment for vasospasm). An opium formulation prepared using the ancient water-dissolving recipe, for instance, contains about 10 percent morphine, .5 percent codeine, and .2 percent thebaine (an opioid that is not itself clinically useful but is the starting point for synthesizing other opioids such as oxycodone). In 1826, a young German pharmacist named Friedrich Sertürner became the first researcher to isolate one of the pure active components of opium. He named the chemical "morphine" after Morpheus, the Greek god of dreams, ushering in the modern era of opiates—and the modern era of opiate abuse.

Commercial production of Sertürner's morphine commenced in 1827 at the Engel-Apotheke ("Angel Pharmacy") in Darmstadt, Germany. The Engel-Apotheke was owned by Emanuel Merck, a descendent of Friedrich Jacob Merck, who founded the German apothecary in 1668. Engel-Apotheke expanded rapidly and eventually became the pharmaceutical company Merck, its rapid growth driven by the strength of its morphine sales. Merck first marketed morphine to the general public as a superior alternative to opium, and soon morphine addiction became even more common than opium addiction.

In 1897, researchers at Bayer Company in Germany used the new science of synthetic chemistry to create a novel chemical variation of morphine that they dubbed "heroin" because it was expected to have heroic effects in treating disease. We now know there is no disease for which heroin is an effective treatment, let alone a heroic one, though Bayer initially marketed heroin directly to the public as a cough suppressant and, absurdly, as a "non-addictive cure for morphine addiction." One nineteenth-century Sears Roebuck catalog peddled a handy heroin kit: one syringe, two needles, two vials of Bayer heroin, and a carrying case—all for the bargain price of $1.50.

It was eventually discovered that the human body metabolizes heroin into several smaller compounds, including morphine, revealing that heroin was not a cure for morphine addiction—it served as a straight substitute for morphine. Yet even though heroin is broken down into morphine, there are important differences between the two compounds. Heroin produces greater psychological stimulation and a much more intense euphoria compared with morphine, and consequently is far more addictive. The morphine addict takes his drug to prevent the appearance of withdrawal symptoms. The heroin habitué, in contrast, takes his drug in order to attain an exulting, blissful high that makes everything bad disappear—at least, until the drug wears off, when everything bad comes rushing back worse than ever. When it became clear that Bayer had actually exacerbated opiate addiction, the company was savaged in the press, marking one of the first public-relations disasters for the modern pharmaceutical industry.

For centuries, exactly how the opiates produced their pain-relieving effects was a great scientific conundrum. Clearly, the poppy had not been guided by the hand of evolution to suppress human coughing or create human addicts. Even in the 1970s, when neuroscience research began to take off, it remained an inscrutable mystery why a Central Asian herbaceous plant held such rapturous power over our brain. Finally, two groups of scientists working independently at the University of Aberdeen in Scotland and Johns Hopkins University in Baltimore solved the neurochemical riddle in 1975.

They discovered that opiates act on specialized receptors in neurons known as endorphin receptors. Eric Simon, one of the discoverers of these receptors, coined the term "endorphin" as an abbreviation of "endogenous morphine," meaning, "morphine produced naturally in the body." Endorphins are naturally occurring hormones produced by the pituitary gland and hypothalamus that produce feelings of well-being and reduce painful sensations. These hormones produce their effects by binding to the endorphin receptors. Humans have nine different types of endorphin receptors, and each opium compound has a distinctive pattern of engaging these nine receptors. This unique pattern of receptor activation determines the characteristic physiological effects of each compound—euphoria, analgesia, sedation, constipation, and so on. When the opiate compound binds with a particular endorphin receptor, the receptor sends a signal into the neuron commanding it to produce other molecular compounds that in turn trigger circuits in the brain that generate the feelings of euphoria and analgesia.

Even when the operation of the opiates on the human nervous system was finally explained, the age-old question remained: why in the world are these brain-jiggering compounds manufactured within a flower? Scientists now have a pretty good answer. Over the eons, most plants have evolved various toxins to protect themselves from being eaten by insects and animals. In response, animals and insects counter-evolved ways to protect themselves from these toxins, such as by degrading them with liver enzymes or developing a blood–brain barrier to prevent toxins from entering their central nervous system. Plant compounds are the product of a relentless arms race between the vegetable and animal kingdoms, a biological death match still ongoing. Scientists speculate that the poppy plant's biochemical pathway for opiate production initially evolved to make neurotoxins that fended off insects.

These botanical opiates were always second-rate toxins, however. They *alter* the behavior of beetles and grubs, sure—but other plants assemble far more efficacious toxins, like strychnine, a poison that induces muscular convulsions and, eventually, asphyxiation. Nevertheless, opiate "toxins" were good enough to protect the poppy from chewing, gnawing bugs to enable the plant to survive all the way into the twenty-first century.

Meanwhile, as poppies were evolving opiates as a way of impairing hostile insects that were sensitive to the toxins, mammals were simultaneously evolving pain-blocking receptors in neurons along a completely independent evolutionary pathway—receptors that by happenstance respond to opiate compounds. Thus, the botanical-chemical system that produces opiates in poppies has absolutely nothing in common with the system that responds to opiates in mammals. In terms of naked statistical probabilities, it is extraordinarily unlikely that a molecular configuration that evolved in plants as a crude insect repellent would also evolve in the sophisticated brains of mammals as a pain-mediator—but, somehow, Mother Nature twice pulled the same chemical volume from the pharmaceutical library of Babel for two disparate missions.

Once our fun-loving Neolithic ancestors stumbled upon the agreeable effects of the milky poppy sap, they began selecting the seeds from those plants that produced the most euphoric intoxication. And today, after many thousands of years of human-guided selection, modern poppy varieties are turbo-charged opiate factories compared to the original species our forebears discovered on the steppes of Central Asia. Studies have shown that even a few generations of selective breeding can dramatically boost the potency of pharmaceutically active compounds in plants. Marijuana is one example. The psychoactive kick of contemporary cannabis plants has been amped up to seven times the potency of the pot smoked at the Woodstock festival in 1969, as measured by the levels of the active ingredient, THC.

The randomness of opium's effect on our brain is underscored by the fact that virtually every compound found in plants has no beneficial effect whatsoever when ingested by humans. Instead, if you swallow a randomly selected leaf or root or berry, most of the time you will get sick. Only about 5 percent of the 300,000 known species of flora are edible. Seventy-five percent of the world's food is generated from twelve species of plants and five species of animals. Yet, prehistoric drug hunters discovered a Vindication in the form of a mind-bending botanical narcotic that became the all-time best-selling medicine in the history of our species. In 2011, more than 130 million prescriptions were written for Vicodin alone, a generic opioid drug derived from codeine, the most prescriptions for *any* drug that year.

Despite the immense commercial success of opiates, the possibility of even greater profits dangles before any drug hunter who might discover a synthetic improvement over Mother Nature's opiates. The ideal painkiller would be: (1) non-addictive, (2) non-sedating, and (3) able to relieve even the most excruciating pain. While opiates are the most high-impact pain relievers available, they are both psychologically and physiologically addictive, they induce drowsiness and constipation, and at not particularly high doses they can halt breathing, causing death. In comparison, NSAID (nonsteroidal anti-inflammatory) pain relievers like aspirin and ibuprofen are neither addictive nor sedating and have virtually zero risk of causing death—an improvement, to be sure, but they do not help with severe or excruciating pain.

When I worked for Wyeth, we had a research group devoted to the development of better pain relievers, a quest shared by all Big Pharma companies. Most of these pain projects focus on blocking some type of ion channel in neurons involved with the transmission of painful stimuli. One of the most interesting lines of inquiry at Wyeth originated with a group of fascinating and ill-fated patients who suffer from an extremely rare condition known as congenital insensitivity to pain (CIP). This condition is caused by mutations in a gene that encodes a voltage-gated sodium channel in neurons known as Nav1.7. Without this ion channel, people cannot feel pain. This might seem like something wonderful, but without the sensation of pain people often injure themselves performing mundane activities, like putting their hands in boiling water or dropping a brick on their foot—acts that feel little different than resting their head on a pillow. In developing countries, people with CIP generally do not long survive, though in the West they can often survive into adulthood if their families have the resources to protect them from inadvertent injury 24/7.

At Wyeth, we realized that if we could somehow mimic the effects of the Nav1.7 ion channel mutation, then we might be able to engineer a drug that could overcome any level of pain, no matter how debilitating. Like everything in drug hunting, this is easier said than done. The painkiller group at Wyeth devoted thousands of man-hours and millions of dollars to the project. Decades later, the Nav1.7 ion channel project has still not produced a single FDA-approved drug, and the dream of a non-addictive, non-sedating, intense-pain-relieving medication remains just that—a wistful dream. As I write this, the best analgesic is still the oldest analgesic.

The existence of high-performance painkillers in the poppy is the result of pure, naked chance, but even the most ardently science-minded of observers cannot help but feel there is something cosmically appropriate in the fact that the most effective mollifier of human agony is found beneath the velvety petals of a cheerful little flower.

2 Countess Chinchón's Cure

The Library of Botanical Medicine

The library of plants

"The plant is hot and has extreme curative healing power. . . . A drink mixed of freshly pressed plant juice with honey and wine fights melancholy, clears the eyesight, strengthens heart and lungs, warms the stomach, cleans the gut, and moves the bowels regularly."

—Hildegard von Bingen on absinthe, in *Physica,* c. 1125 AD

There have always been two distinct breeds of physicians. The *practitioners,* such as primary care doctors and brain surgeons, focus on providing effective care to their patients. The *researchers,* in contrast, seek out new medical discoveries that may benefit many. These days, the most prevalent form of the medical researcher is the physician–molecular biologist, typically an MD-PhD hunting for new cures within genomic science. But from the Renaissance back into the murky depths of antiquity the most common type of medical researcher was the physician-botanist. Why? Because virtually all new drugs were found within the chlorophyll kingdom of plants.

Pharmacology was essentially a specialized branch of botany for the first ten thousand years of human civilization. We might call this era of drug hunting the Age of Plants. All the variegated specimens of the vegetative world—flowers, roots, seeds, bark, sap, moss, seaweed—were considered God's own pharmacopeia, to be harvested and husked and milled and boiled into beneficial tonics.

(Indeed, the English word "drug" comes from an ancient French word, "drogue," which referred to dried herbs.) The discovery of new balms required expertise in both human disease and plant lore, and thus nearly every pharmaceutical revelation from the dawn of history until the eighteenth century was perpetrated by a physician-botanist. Perhaps the most esteemed of these early botanical drug hunters was a German prodigy by the name of Valerius Cordus.

Born in Hesse, Germany, in 1515, Cordus was the son of a physician and the nephew of an apothecary. His uncle took the young Cordus on drug-hunting expeditions in the wilds of northern Germany, where they gathered up medicinal plants, and then revealed to him the arcane methods for distilling their horticultural bounty into potions and ointments. Cordus came of age during a time when most apothecaries had an alchemical bent, when elixirs for enchanted romance were just as common as powders for crotch rash. But from the moment he started university in the scholarly city of Wittenberg, Cordus exhibited no interest in superstition or interpretative divination. Instead, he insisted that the apothecary craft should consist solely of careful observation and verifiable results.

While still a graduate student, Cordus began delivering sophisticated lectures on a famed ancient Greek apothecary named Dioscorides, a physician-botanist who lived around 50 AD who penned a five-volume encyclopedia about herbal medicine known as *De Materia Medica*. This hefty pharmacopeia detailed everything that was known in the ancient world about medicinal substances. It described nearly one thousand different drugs. Dioscorides's tome had served as Europe's Physician's Desk Reference for more than fifteen hundred years, an astonishing run that did not rest on the accuracy or clarity of the *De Materia Medica* but rather on the fact that no serious effort was made to improve upon it.

Cordus's lectures on Dioscorides were so highly regarded that even professors attended them—a rarity at the time and even more impressive considering that Cordus was barely out of his teens. Though Cordus praised *De Materia Medica,* he also suggested that it was high time for Europeans to break from the stodgy old antiquarians and develop their own modern manual of medicines. To fulfill this new mission, Cordus devoted himself to two tasks after he left the university. He searched the world for new plants that could be the source of new drugs. And he began writing a new pharmacopeia based on evidence rather than tradition.

In 1543, at the youthful age of twenty-eight, he published the *Dispensatorium.* This landmark opus was the first major pharmacological document to exclude all reference to the supernatural and mystical, focusing exclusively on empirical knowledge about the properties and preparations of plants. It listed more than 225 medicinal plants, including myrrh, crocus, cinnamon, piperis, absinthe, gum Arabic, calamus, camphor, cardamom, cucumeris, citrulli, margaritarum, roses, anise, and balsam. Because of its careful observations on such a wide variety of flora, the *Dispensatorium*'s contribution to scientific botany became just as important as its contribution to scientific pharmacology. Cordus's radical new pharmacopeia became the most widely used apothecary manual for the next century.

But Cordus was not satisfied with documenting all that was already known about drugs. He was obsessed with discovering new ones, too. Influenced by his childhood expeditions with his uncle, he voyaged to exotic and obscure locales in the hope of unearthing new plants to add to his growing compendium of medicines. He also began experimenting with chemistry, a fledgling field that was still closer kin to occult alchemy than verifiable science. Once again, Cordus distinguished himself through meticulous observation, recording only those results that could be replicated.

As a drug hunter, Cordus mostly scavenged through the library of plants for his Vindications. But he was also a formulator, attempting to contrive new versions of drugs using techniques from the nascent field of scientific chemistry. Cordus's greatest success was a medicine that is still in use today in a handful of developing nations—ether. While Cordus was not the first person to discover ether, which he referred to as "sulphur" or "vitriol," there is no question that Cordus was the first to provide a reliable account of its synthesis from sulfuric acid and grain alcohol. He systematically described the chemical properties of both "sour oil of vitriol" and "sweet oil of vitriol" (the latter of which eventually became modern ether), including its high volatility and its unfortunate tendency to blow up in a fiery explosion. Like all of his research, however, his investigation of ether was ultimately directed towards the therapeutic. He wrote detailed reports of the medicinal applications of *oleum dulce vitrioli,* including its promotion of mucus secretions and its amelioration of hacking coughs. We will return to ether in the next chapter, where we will see how the drug was almost single-handedly responsible for the establishment of the modern pharmaceutical industry.

So what was life like for a Renaissance drug hunter? Sadly, it could be tragic and short. In the summer of 1544, Cordus ventured into the mosquito-infested swamps of Florence and Pisa, boldly prospecting for new botanical varietals in the sludge. After returning to Rome with his harvest, he was stricken with malaria and perished—a victim of his own drug-hunting ambitions. He was twenty-nine years old. At the time of his death, he had directly contributed to the foundations of no less than three fields of science: botany, chemistry, and pharmacology. His epitaph reads: "Valerius Cordus, while still a youth, explained to men the working of Nature and the powers of plants."

As Europeans began to colonize the New World in the wake of Columbus's voyages of exploration, botanical drug hunters extended their quest for exotic plant tinctures to the uncharted lands on the far side of the world. One of the most important discoveries was the bark of a tree found in the western jungles of Bolivia and Peru, a tree we now call *chinchona*. The aboriginal Quechua people brewed its bark into an earthy, bitter tea they imbibed to prevent malaria. The Spanish conquistadors quickly adopted the wondrous bark as their own; one Augustinian monk named Calancha wrote in 1633: "A tree grows which they call 'the fever tree' in the country of Loxa, whose bark, of the color of cinnamon, made into a powder amounting to the weight of two small silver coins and given as a beverage, cures the fevers and tertians; it has produced miraculous results in Lima."

In the fifteenth century, *tertian* was a term used to describe a fever that was intermittent, rising and falling—the type most commonly seen in malaria. The reason that fevers come and go in individuals suffering from malaria is because the parasite that causes the disease replicates in synchronized waves inside the host's red blood cells. After a round of replication concludes, the red blood cells burst open, and all the parasites simultaneously dash out to invade new blood cells. This process induces fever when chemical fragments from the burst cells enter the bloodstream (they are toxic products from the degradation of hemoglobin). Once the parasites successfully penetrate a new population of red blood cells, the fever resolves, and a new cycle of infection commences.

One story holds that chinchona bark was used to treat Countess Anna del Chinchón, the wife of the viceroy to Peru, in 1638. (The genus of plants that produce quinine were named in her honor by Carl Linnaeus, the "father of modern taxonomy," since he believed she was among the first Europeans cured by the bark.)

Her supposedly miraculous recovery led to the introduction of chinchona into Spain in 1639 as a malaria treatment, and for years the bark was called "los Polvos de al Condesa"—the Countess's Powder. It is true that the viceroy did bring a large quantity of chinchona to Spain, but what is not clear is whether his wife was ever treated with "los Polvos de al Condesa" or if the epithet was merely a marketing ploy invented by the viceroy to help promote the sales of his ample stash of bark.

The Jesuit missionaries in South America quickly established themselves as the main importers and distributors of chinchona in Europe, where it was often called "Jesuit's bark." It soon became one of the most valuable commodities shipped from Peru to the Old World. This New World drug, however, was not without its controversies.

Traditional physicians of the era, known as Dogmatists, did not believe in the bark's powers of healing because it did not conform to the teachings of the ancient physician Galen and his theory of the four bodily humors, which held that malaria should be cured through purges (usually a forceful evacuation of the bowels). The Dogmatists were opposed by the Empirics, early rationalists who believed that medical remedies should be sought out through observation and experimentation. This debate raged across Europe for decades and produced a storm of claims and counterclaims about the American bark. Many charlatans and hucksters took advantage of this climate of pharmaceutical uncertainty, the most famous of whom was an English apothecary named Robert Talbor.

Talbor promoted his own remedy for malaria. In 1672, he published "Pyretologia, A Rational Account of the Cause and Cure of Agues," a compact volume which—although scientific in appearance—was basically a marketing brochure touting his wonder drug. Though he described the method of *administering* Pyretologia in great detail, all he reported of its composition was that it was "a preparation of four vegetables, whereof two are foreign and the other domestick." As he hawked his own remedy, he vociferously warned against the use of chinchona bark:

> And let me advise the world beware of all palliative Cures and especially that known by the name of Jesuits' Powder, as it is given by unskillful hands for I have seen dangerous effects follow the taking of the Medicine uncorrected and unprepared.

Talbor was a man motivated by lucre. When physicians solicited him for a more complete account of his mystery balm, Talbor wrote that prior to revealing its ingredients he deserved to be compensated for his efforts:

> I intend hereafter to publish a larger, and fuller account of my particular method, and medicine, not being willing to conceal such useful remedies from the world any longer, than till I have made some little advantage myself, repay that charge and trouble I have at in the search and study of so great and unheard of secrets.

He eventually earned the payoff he was looking for by curing the son of Louis XIV using Pyretologia. The French Sun King rewarded Talbor with "3,000 gold crowns and a lifetime pension." Nevertheless, even though Talbor was frequently called upon to disclose the ingredients of his remedy, he never divulged his secret recipe. A year after Talbor's death, several apothecaries finally identified the key component of Pyretologia: the bark of the chinchona tree.

Two more centuries passed before the active chemical in chinchona was finally isolated in 1820 by two French apothecaries, who dubbed it *quinine*. The compound had a transformative impact on human civilization. It opened up malaria-ridden lands across the globe to Western colonization, including huge swaths of South America, North America, Africa, and the Indian subcontinent that were previously too dangerous to inhabit. European colonists' frequent consumption of quinine also gave rise to a new alcoholic cocktail that remains popular to this day—the gin and tonic. The typical nineteenth-century imperial British bureaucrat, reclining on a veranda wreathed with mosquito nets in some remote outpost of the British Empire, ordered gin and tonics from his native servants that he sipped as he enjoyed the setting sun. The tonic water contained the quinine, but its bitter taste was difficult to get down, so gin was added to mask the flavor. (If adding hefty swigs of strong grain alcohol was considered an improvement, you can probably guess just how unpleasant quinine actually tastes.) In addition, quinine has poor solubility in water, so mixing it with alcohol made it easier to dissolve the medicine.

Quinine was one of the last great drugs to be discovered during the Age of Plants. The Spanish physician-botanist Nicolas Monardes published a lengthy treatise in 1574—translated into English as *The three books of joyfull newes out of the newe founde worlde*—that describes more than one hundred other New World plants that could be used as medicine. This list included curare, coca (cocaine, used by the indigenes for treating hematomas and eventually prescribed by European doctors for a variety of ailments), cacao (chocolate, used for treating depression and exhaustion), sassafras (used as a very ineffective treatment for fevers, including syphilis), arbor vitae (the "tree of life," used to treat scurvy), tobacco (used to treat a variety of maladies), snakeroot, smilax, maiden hair fern, rose mallow, guaiacum (for the pox), various purgative nuts, oil from the fig tree of hell (a purgative), ipecacuanha (another purgative), canafistula, *estoraque,* (American) balsam (used to treat many ailments), and white jalap. Out of this list, the only plants still used in scientific medicine today are quinine, curare (as a paralytic in certain forms of surgery), and ipecacuanha (to induce vomiting). Chocolate, of course, is sometimes considered an aphrodisiacal drug—and is sometimes used for self-medicating during depression—but is no longer found on the pharmacist's shelf.

In many ways, Valerius Cordus's short life marked one of the most momentous turning points in the quest for medicine, since his career embodied the transition from searching through the library of plants to searching through the next major pharmaceutical library, the library of synthetic chemistry. His tragic death from overzealous prowling in the boggy wilderness signaled the end of the most enduring era of drug hunting.

Today it is exceedingly rare to discover new medicines from plants, because the world's botanical largess has been so exhaustively harvested, husked, and scrutinized. In the 1990s, for example, I was working for the pharma company Cyanamid when our drug development team decided to forage through exotic foliage around the world in the hope of exhuming some new horticultural medicines. This would have required us to work with an expert botanist, but by the late twentieth century botany had devolved into a minor scientific discipline that was no longer pursued with much vigor by American universities. Consequently, we could not find anyone with the knowledge and interest to assist us with the project. (It might seem strange that scientific expertise could be so easily lost, but the diminishment of previously robust fields happens all the time. When I was a graduate student at Princeton, a scientist visited our biology department requesting access to our collection of bivalves, mollusks with two shells like clams and oysters. Nobody knew anything about the collection. The department chairman made some inquiries and learned from the staff that, during a remodeling effort ten years before, one of the workers discovered a bunch of seashells and threw them out. There was no outcry at the time, because there was no longer any professional interest in mollusks. It turned out that the Princeton bivalve collection had been considered one of the best in North America.)

Since we could not track down a suitable botanist in the United States, we established a collaboration with the Institute of Cell Biology and Genetic Engineering in Kiev, Ukraine, which still had a very active program of botanical research. On our behalf, the institute sent botanical expeditions to remote regions across the globe, including the former Soviet Union (Ukraine, Russia, Kazakhstan, Azerbaijan, Kyrgyzstan, Uzbekistan), South America, Africa (Namibia, South Africa, Ghana), and Asia (China and Papua-New Guinea). The Kiev botanists collected some fifteen thousand species of plants. Despite this impressive haul of obscure herbs, shrubs, and flowers, the Cyanamid team failed to find a single new useful compound. After millennia of human exploitation, the library of plants might just be emptied of its Vindications.

3 Standard Oil and Standard Ether

The Library of Industrial Medicine

An early pharmaceutical factory

"I have seen something today that will go around the world."

Dr. Henry J. Bigelow, 1846

Though the Age of Plants was the longest and most prolific era of drug hunting, botany was becoming overshadowed in the earliest years of the Renaissance by the rise of alchemy, which might be more accurately described as the rise of pre-scientific chemistry. The most noble—and, potentially, the most lucrative—aspiration for any medieval alchemist was the "philosopher's stone," the term of art for any method of transforming a base element, like lead, into a precious metal, like gold. One typical formulation can be found in an alchemical manuscript discovered in a Jewish synagogue in Old Cairo dating from the twelfth century: "Combine mercury, horse manure, pearl, white alum, sulfur, clay mixed with hair and a couple of eggs and you will obtain good silver, God willing." Today, we know that the most crucial step of this recipe—"God willing"—requires either nuclear fission or nuclear fusion, technologies unavailable in a culture that had no conception of the atom. Horse manure, on the other hand, remains a common ingredient of misconceptions to this very day.

Any discipline that relies on the agglomeration of dung and divine intervention is unlikely to produce useful innovations, and for many uneventful centuries from the 1100s through the 1600s the alchemical drug hunters offered precious little in the way of practical advances in pharmacology, though they added a great many drug formulations that were barely useful at best, and lethal at worst. Valerius Cordus finally shrugged off the hazy bonds of the occult in favor of scientific observation. His reliable recipe for "the oil of vitriol" proved to be far more transformative than the misbegotten quest for the philosopher's stone.

The Swiss-German alchemical formulator Paracelsus, a contemporary of Cordus, wrote that ether would put chickens to sleep for "a moderate time" without hurting them, though Paracelsus failed to consider using it to put humans to sleep. Likewise, though Cordus patiently recorded several medicinal uses for ether based on his thoughtful experimentation, there is no record that he was aware of its effects as an anesthetic. Cordus's formulation of ether remained a standard if minor part of the physician's pharmacopeia for the next three centuries, used as a chemical solvent and as a (presumably extremely ineffective) treatment for headaches, vertigo, epilepsy, palsy, hysteria, rheumatism, and many other diseases. But even the most forward-thinking physicians of the early nineteenth century exhibited no greater imagination than the medieval apothecaries regarding the oil of vitriol's uses.

In 1812, one recommended use for ether appeared on the very first page of the very first volume of the *New England Journal of Medicine*. Dr. John Warren, one of the founders of Harvard Medical School and one of the most prominent physicians of the time, wrote an article on the treatment of angina, a painful condition that feels like your chest is being squeezed. Today we know angina is caused by insufficient supply of oxygen to the heart, but since

Warren lacked adequate knowledge of the disease he offered up a rather dubious catalog of putative treatments: bathing the feet in warm water, bleeding, nitrate of silver, fetid gums, tobacco smoking, opium, and, finally, ether.

Not only was ether recommended as an angina treatment, by 1830 it was best known to the public as a recreational intoxicant at giggly parties known as "ether frolics," where wealthy straitlaced Victorians huffed the vapors of the oil of vitriol and proceeded to flop around, stumble into the furniture, or pass out entirely. Ether was also prescribed then as an antiseptic, a cleansing solvent, an expectorant in cough medicines, a carminative (that is, an anti-flatulent), and—rather improbably—a stimulant in cases of fainting, where it was sometimes combined with the far more efficacious aromatic spirits of ammonia. There was one medical use, however, for which ether had never been prescribed in all its years of existence.

Surgery was uncommon prior to the middle of the nineteenth century. For one thing, surgery was very dangerous. Infection was an almost-inevitable consequence of any surgical procedure, and these infections were often lethal. Aseptic techniques were not practiced at all prior to the establishment of the germ theory of disease in the late nineteenth century. Worse, knowledge of disease pathways was either rudimentary or entirely nonexistent, and as a result there was no consistent scientific rationale for surgical intervention. Finally, surgery was conducted without any anesthesia, and as a result, was agonizingly, excruciatingly, soul-wrenchingly painful.

It's hard for us to imagine what surgery must have been like before the use of anesthesia, though we can get some idea from George Wilson, a prominent professor of medicine who had had his foot amputated in 1843 and described the unspeakable awfulness of the procedure:

> The horror of great darkness, and the sense of desertion by God and man, bordering close on despair, which swept through my mind and overwhelmed my heart, I can never forget, however gladly I would do so. During the operation, in spite of the pain it occasioned, my senses were preternaturally acute, as I have been told they generally are in patients in such circumstances. I still recall with unwelcome vividness the spreading out of the instruments: the twisting of the tourniquet: the first incision: the fingering of the sawed bone: the sponge pressed on the flap: the tying of the blood vessels: the stitching of the skin: the bloody dismembered limb lying on the floor.

In the first half of the nineteenth century, surgery was an emergency procedure—the amputation of a limb to prevent fatal gangrene, the drainage of an infected abscess, or a cystotomy for an excruciating bladder stone (one of the few maladies considered more painful than the surgery itself). Fine dissection and careful technique were simply not possible because patients twisted and contorted in pain under the surgeon's blade. The best strategy for a successful operation was *speed*. The faster the procedure was completed, the less the excruciating pain, and the less convulsive the patient.

Spectators in early nineteenth-century operating galleries pulled out their pocket watches to time the pace of surgical procedures. Dr. Robert Liston, for instance, a Scottish surgeon who practiced at University College Hospital in London, was famed for the swiftness of his technique. Once, in his haste to amputate a leg, he sliced off the patient's testicles as well. During another brisk amputation of a leg, Liston managed to spare the patient's testicles—but accidentally removed two of his young assistant's fingers. Both patient and assistant eventually died of gangrene, while a spectator watching the same operation died of shock upon seeing Liston's whirling blade slash through his coat, believing that he had just been fatally stabbed. Such were the dangers of surgery in the pre-anesthetic era.

Given the pressing need to relieve surgical pain, physicians experimented with many potential anesthetics. Alcohol, hashish, and opium were all sampled and found wanting. Though they dulled the senses somewhat, they proved to be inadequate for the kind of agony induced by a knife cutting through deep muscles. Physical methods, such as packing a limb in ice or rendering it numb with a tourniquet, were also insufficient to the task. Pain always seared through. Some bolder surgeons went so far as to produce unconsciousness in their patients through strangulation or a severe blow to the head, though most physicians expressed doubts that the benefits of these procedures outweighed their disadvantages. Nineteenth-century surgeons were trained to expect surgery to be a bloody affair, full of writhing and screams, a task that must be completed as hastily as possible. That might be why it took a non-surgeon to imagine the possibility of surgery without pain—a Boston dentist by the name of William T. G. Morton.

In 1843, at age twenty-four, Morton married Elizabeth Whitman, the niece of a former congressman. Elizabeth's prominent and patrician parents objected to Morton's profession. At the time, a dentist was regarded as little better than a barber. The Whitmans agreed to allow their daughter to marry Morton only after he promised to study the far more respectable profession of medicine.

In the autumn of 1844, Morton dutifully enrolled in Harvard Medical School, where he attended the chemistry lectures of Dr. Charles T. Jackson. Jackson was well acquainted with the pharmacological properties of ether, including its anesthetic qualities, but even though he was a bright working physician he apparently never gave serious consideration to the possibility that ether could be applied during surgery. Morton learned about ether during one of Jackson's lectures and, intrigued by its unparalleled ability to put people to sleep, he experimented with the drug on his pet dog, writing:

> In the spring of 1846, I tried an experiment upon a water spaniel, inserting his head in a jar having sulphuric ether at the bottom. . . . After breathing the vapor for some time, the dog completely wilted down in my hands. I then removed the jar. In about three minutes he aroused, yelled loudly, and sprung some ten feet, into a pond of water.

Morton also tested ether on a hen and some goldfish. They, too, softly succumbed. Emboldened by these successes, Morton inhaled the sweet-smelling vapor himself. He passed out, then recovered fully without any noticeable ill effects. Morton finally felt it was time to try it on an actual patient. Morton performed the world's first painless dental extraction in his Boston office, removing an ulcerous tooth from a grateful merchant who history records as Mr. Eben Frost:

> Toward evening, a man . . . came in, suffering great pain and wishing to have a tooth extracted. He was afraid of the operation and asked if he could be mesmerized. I told him I had something better, and saturating my handkerchief, gave it to him to inhale. He became unconscious almost immediately. It was dark, and Dr. Hayden held the lamp, while I extracted a firmly rooted bicuspid tooth. There was not much alteration in the pulse, and no relaxation of the muscles. He recovered in a minute, and knew nothing of what had been done to him.

On October 1, 1846, the *Boston Daily Journal* published an account of Morton's curious experimental procedure. The story made its way to Henry Bigelow, a junior surgeon affiliated with Harvard

Medical School. Intrigued, Bigelow convinced the eminent head surgeon of Massachusetts General Hospital to stage a public test of Morton's purported anesthetic procedure. This was the big time. It was the nineteenth-century medical equivalent to landing a spot on *American Idol*. Mass General was one of the most respected hospitals in the country, and its head surgeon, sixty-eight-year-old John Collins Warren, had a national reputation. Warren had previously served as the dean of the Harvard Medical School, an institution his father helped found, and helped establish the *New England Journal of Medicine*.

With the stakes suddenly so high, Morton knew he would be taking a huge risk. It was one thing to fool around with ether in the relative obscurity of a dental office. After all, nobody expected very much from the uncouth and unruly pseudo-profession of dentistry. It was quite another thing to test the drug's performance during life-risking surgery in front of the Brahmin cream of the medical establishment. More than fifty skeptical spectators, including many of America's top surgeons, gathered in the Mass General operating theater on October 16, 1846. Some were genuinely curious about ether's effects. Most, though, expected to watch the public exposure of a charlatan.

The patient, one Edward Gilbert Abbott, suffered from a large tumor bulging out of his neck. Its removal would be a harrowing and excruciating experience—at least, that was how it usually went. Two strong orderlies stood nearby, ready to assume their usual role of holding down a flailing, shrieking patient. But would this time be different?

As the audience watched from the high tiers of seats, the patient was rolled into the operating theater. Warren stood by, waiting. The clock ticked past the appointed time for the commencement of the surgery, but Morton did not appear. Warren turned to the gallery. "As Dr. Morton has not arrived, I presume that he is otherwise engaged." The patient gritted his teeth. The surgeon lifted his scalpel.

Suddenly, Morton strode into the theater. There was good reason for his delay. Since nobody had ever before administered ether during surgery, there was no mechanism for delivering ether vapor in a controlled fashion. Morton had been busy constructing an ingenious apparatus: a round-bottomed chemistry flask that contained sponges soaked in ether. Two ports were attached to the flask with brass fittings, and, via a clever system of leather flaps, air was drawn from one aperture over the ether-soaked sponges and inhaled through the other.

Warren stepped back and said, "Well, sir, your patient is ready." Surrounded by a silent and largely unsympathetic gallery, Morton went to work administering the ether from his newly contrived glasswork. The patient breathed a few slow breaths of the vapor and his eyes gently shut. Morton turned to the surgeon. "Dr. Warren, *your* patient is ready."

The operation began. As the scalpel sliced deep into his neck, the patient showed no reaction. Even so, the slow heaving of his chest demonstrated he was clearly alive and breathing. The audience gaped in awe. Today, we take the existence of anesthesia for granted, but imagine just how wizardly this must have seemed to a physician at the time—there was some magical *substance* that somehow completely shut the mind off from all awareness, yet left the body's physiological operations unaffected. It was a moment as revolutionary to medicine as the discovery of gunpowder was to warfare or the invention of powered flight was to transportation. When the operation was completed, Dr. Warren turned to the audience and announced, "Gentlemen, this is no humbug."

As word of the breakthrough spread, ether was rapidly adopted as an essential component of every major surgery, creating an ever-increasing demand for the compound. There was one big obstacle to filling this demand, however. Ether was quite tricky to make. It required advanced chemistry techniques that lay well outside the expertise of the apothecary.

Since ancient times, apothecaries were the place to go to obtain remedies. The apothecary was typically a small, local shop or stand run by a single proprietor. In the seventeenth century, apothecaries in Europe were the first to become formally organized. In London, the Worshipful Society of Apothecaries was granted a royal charter by King James I in 1617 as a professional organization focusing on the formulation of medicines. Medicines were not their only product, however. Apothecaries also sold spices, perfumes, honey, dyes, saltpeter (an ingredient in both medicine and gunpowder), camphor, gum benjamin (a wood resin used as incense, flavoring, and medicine), frankincense, aniseed, capers, and molasses as well as items that seem more suitable for a witch's cauldron than a physician's cabinet: hearts of stags, frog spawn, crayfish eyes, bulls' penises, the flesh of vipers, swallow nests, and the oil of foxes. In *Romeo and Juliet*, Shakespeare famously describes an apothecary's shop in Renaissance Italy:

> And in his needy shop a tortoise hung,
> An Alligator stuff'd and other skins
> Of ill-shaped fishes; and about his shelves
> A beggarly account of empty boxes.

By the seventeenth century, the apothecary had become increasingly specialized in the art of making drugs, and a would-be apothecary required a long and arduous period of apprenticeship before becoming a credentialed professional. Apprenticeships lasted seven years, and apprentices were required to participate in frequent "herbarizing" expeditions, gathering botanical samples in the wild in order to gain sufficient familiarity with medicinal plants. To become an apprentice, one had to demonstrate knowledge of Latin, the international language of pharmacology, and in England a successful candidate had to satisfy the Society of Apothecaries that "he had knowledge and Election of Simples [plants or herbs used for medicinal purposes] and was able to prepare, dispense, handle, commix and compound medicines." Distinctly absent from the training of an apothecary was any education in the young but rapidly expanding field of chemistry.

At the time of Morton's public demonstration of ether, American apothecaries operated small retail shops that served their local community. Drugs were prepared by the apothecaries according to their own individual interpretation of common recipes, many of which dated back to Cordus's three-hundred-year-old *Dispensatorium*. Thus, a formulation of opium you bought from a pharmacist in New York could differ greatly from a formulation of opium purchased from a pharmacist in South Carolina. On top of this already-significant variability in the composition of basic drugs, ether was particularly difficult to synthesize, for it required intricate knowledge of organic chemistry and chemical purification procedures that were well beyond the means of most apothecaries. As a result, surgeons found that they needed to procure ether from the nascent industry of chemical suppliers rather than risk the wildly unpredictable (and frequently unavailable) versions of ether offered by apothecaries.

Unfortunately, surgeons soon learned that the ether from chemical suppliers was not very reliable either. A batch bought from a chemical supplier one day might have a totally different purity than a batch bought from the same supplier a month later. Even worse, there were vast differences between the ether offered by different suppliers, with the most incompetently concocted versions failing at their most important task—putting patients to sleep. This lack of consistency made it very difficult to know how much ether to administer to ensure patients stayed unconscious without actually halting their respiration and killing them. Surgeons needed a standardized formulation of ether they could trust.

The demand for standardized products was felt across many industries at the dawn of the Industrial Age in the mid-nineteenth century. Before the invention of electricity, the entire nation was illuminated by kerosene lamps. The largest and most successful corporation in the history of the world, Standard Oil, attained success because it was the very first company to standardize its preparation of kerosene—hence the company's name. If you bought a gallon of Standard Oil's kerosene in California, it was exactly the same as a gallon of Standard Oil's kerosene purchased in New York. Rockefeller used standardization to outcompete hundreds of other local kerosene manufacturers and eventually establish a monopoly of the entire energy market, all because he offered a reliable, consistent product that customers could count on.

As the demand for ether skyrocketed in the 1850s, apothecaries were simply not equipped to provide the kind of mass-produced standardized ether products that hospitals and surgeons hungered for. But like Rockefeller when he figured out how to standardize kerosene, another enterprising man from humble beginnings established an entire industry by figuring out how to standardize ether.

Edward Robinson Squibb was born a Quaker in Wilmington, Delaware in 1819. Squibb graduated from Jefferson Medical College in Philadelphia, Pennsylvania in 1845 at the age of twenty-six, just one year prior to Morton's ether demonstration, then joined the U.S. Navy as a ship's physician. Squibb spent four years with the Atlantic and Mediterranean squadrons, where he became increasingly concerned about the Navy's poor treatment of the men under his care. He published critical accounts describing inadequate diet, frequent floggings, and—most significantly—the poor quality of the medicine dispensed aboard Navy ships.

Squibb's grievances reached the Navy's Bureau of Medicine and Surgery. They responded to his complaints by ordering him to establish a Naval Laboratory in the Brooklyn Navy shipyard with a mission of producing high-quality drugs. One of his first tasks was to evaluate the myriad brands of ether. Squibb took a six-month leave to attend refresher courses at Jefferson Medical School, where he studied chemical synthesis techniques in order to better understand the manufacture and evaluation of ether. When he returned to the Naval Laboratory, Squibb tested the different commercial formulations of ether and found them to be fantastically variable in purity. He decided to develop a method for producing ether of a consistent quality and quickly discovered what a technical challenge it truly was.

Ether is highly flammable and highly explosive, but the process for synthesizing ether requires both heat and flame. During one of Squibb's early experiments, an explosion burned off both of his eyelids, and for the rest of his life he had to place a dark cloth over his eyes at night in order to sleep. But in 1854, the persistent physician-chemist achieved a manufacturing breakthrough. He dramatically improved the process of ether production by replacing the open flame with steam passing through a coil.

When budget cuts forced the Brooklyn Naval Laboratory to close in 1857, Squibb decided to start his own company around his new method. He founded the first American pharmaceutical manufacturing factory at a site adjacent to the Brooklyn Navy Yard and named his new company E. R. Squibb and Sons. The American Civil War created enormous demand for medical supplies, and Squibb's contacts in the Navy put him in an excellent position to secure military contracts. The physical location of the company was also favorable; Squibb could simply walk across the street to the Navy Yard to negotiate contracts, then drive a wagon across the same street to deliver the purchased products.

When the war ended, Squibb's success continued to grow. The company's reputation for producing trustworthy, standardized medicines led to high national demand for Squibb products. This consistency is embodied in Squibb's original logo, used all the way into the 1980s when the company was acquired by Bristol-Myers. The logo comprised a marble pediment emblazoned with the word "reliability," supported by three columns with the words "uniformity," "purity," and "efficacy."

Consider how different Squibb's business model was from that of our own pharmaceutical industry. Squibb did not offer original or unique drugs. Instead, it outcompeted other suppliers by manufacturing more *consistent* drugs. Today, drug makers do not compete on reliability or consistency, since modern consumers presume that any drug they find on the shelf is going to be perfectly standardized. (Can you imagine a customer's puzzled reaction to a television ad that proudly attested, "Every bottle of Tylenol is the same!"?) During the Age of Plants, the drug-making industry was like community theater, each apothecary serving his local neighborhood by formulating drugs according to his own personal tastes and inclinations. But now Squibb began making the pharmaceutical equivalent of the Hollywood blockbuster—formulaic, big-budget productions marketed to the entire world. Big Pharma was born.

A little more than a century after Squibb began producing

ether, I landed my first industrial R&D job with E. R. Squibb and Sons. Although unrecognizable as the company whose Brooklyn factory had once been the site of violent ether explosions—the modern-day Squibb had acquired perfume and candy businesses, among many others—it had nevertheless managed to preserve much of its founder's abiding philosophy regarding which employees were most important. E. R. Squibb, a physician himself, believed that the physicians and biologists should direct the development of new medicine formulations, while the chemists should merely serve a supporting role.

I did not fully appreciate Squibb's medicine-first culture until I had worked for two other pharmaceutical companies. One of these was Cyanamid, which at its heart was a chemical company rather than a medicine company. American Cyanamid was originally founded in 1907 to produce a basic fertilizer ingredient known as calcium cyanamid, then grew by looking for new ways to exploit its newfound expertise in chemistry. Shulton, its consumer division, developed cleaning and grooming products, such as Old Spice aftershave, Breck shampoo, Pine-Sol cleaner, and Combat roach traps. Its agricultural division made chemical pesticides. Its chemicals division made industrial chemicals. In every division, including its pharmaceutical division, Lederle, chemicals and chemistry came first. It was a professional shock to me as a molecular biologist when I was demoted, in a sense, from the "A" team at Squibb to the "B" team at American Cyanamid.

But I received an even more important lesson about the influence of a business's attitude toward its drug hunters when I unexpectedly started working for American Home Products (AHP) in the late 1990s. I abruptly became an AHP employee after it bought the pharma company where I was initially employed. AHP was a financially driven holding company, which meant that the people running AHP would buy any company in any industry if they thought they could somehow squeeze a few drops of profit out of it. If AHP had the choice between earning $10.00 an hour for shoveling manure or $9.99 an hour for sniffing flowers, they would not hesitate an instant before picking up the shovel. As with most holding companies, there was little apparent rhyme or reason to its esoteric collection of businesses; AHP sold everything from perfumes to sauce pans to Chef Boyardee to vitamins and medicines. And because all AHP executives were directed to focus on the bottom line, any capital expenditure within any division of AHP—even for sums as little as $5,000—had to be reviewed by the corporate finance committee and approved by the CEO, Jack Stafford.

Drug discovery requires a sustained effort that usually takes more than a decade to produce a useful medicine. A corporate focus on short-term profits often has a stifling effect on pharmacological research. Many of my fellow drug hunters at AHP tried to work around the restrictive constraints of the company's capital expenditure policy, most commonly by gaming the system. Pharma scientists would grossly overstate their budgetary needs so that they would have enough funds to continue their research when the inevitable cost cuts came down. My own strategy—at least, at first—was to try to reason with AHP executives and explain how difficult it was to find new drugs when the financial decisions were always made for near-term impact instead of long-term value. Gradually, I realized that there was little hope of changing the minds of managers who were immersed in a corporate culture based on immediate financial calculations rather than one based on the patient and deliberate development of new medicines. While I was working there, I don't think AHP developed a single drug that made a meaningful difference to patients or to medical practice.

It is worth taking a moment to retrace the unlikely path that led to the establishment of the pharmaceutical industry in America, which today includes corporate cultures that are distinctly hostile toward the risk-laden realities of modern drug hunting. Ether was discovered during the very height of pseudoscientific alchemy by a physician-botanist who suggested it be used to treat coughs. Three centuries later, in the early 1800s, it was prescribed for an unwieldy hodgepodge of ailments, though we now know that it is worthless as a treatment for most, if not all, of these maladies. Then, in an attempt to impress his snooty in-laws, a dentist decided to try using this party drug to painlessly remove a patient's tooth and ended up transforming surgery from a shriek-filled horror show to a calm and meticulous craft. And yet, though it revolutionized surgery, ether would not have revolutionized the pharmaceutical industry if it had been easy to make. Since ether required expansive and expensive technologies to produce a standardized compound, it led drug-making out of the apothecary shop and into the factory.

Squibb's success served notice that it was possible to manufacture important drugs on a massive scale. The Age of Industrial Formulation was not about inventing new drugs—it was about finding new formulations for existing drugs that could leverage the rapidly growing young science of chemistry and the new manufacturing techniques of industrial factories to produce standardized drugs on a massive scale. Drug hunters from this era, like Squibb, rummaged through the library of industrial formulations for new recipes for established drugs with a pre-existing market. Some other industrialized formulations included chloroform, morphine, quinine, ergot, jalap (a cathartic that accelerates defecation), ignatia (believed to be a kind of antidepressant), conium (used to treat trembling and palsy), guarana (used like caffeine), erythroxylon (fluid extract of cocaine), and alum (used to constrict tissue, reduce bleeding, or sometimes to induce vomiting).

But this focus on improving the manufacture of existing drugs would soon change. A very different breed of drug hunter arose who searched for their Vindications in the vast new library of molecules known as synthetic chemistry.

4 Indigo, Crimson, and Violet

The Library of Synthetic Medicine

The original bottle of Bayer Aspirin

"The product has no value."

—Heinrich Dreser, director of Bayer Pharmaceutical Research, on aspirin, 1897

If this evening you went looking for the pharmaceutical industries of Switzerland and Germany, you would find their largest and most venerable companies along a single river, the Rhine. The headquarters of Novartis, Bayer, Merck KGaA, Hoffmann-La Roche, Boehringer Ingelheim, and Hoechst all sprawl along the banks of a waterway that flows through the heart of Germany on its windy way to the North Sea. In the 1990s, I learned the reason for this geographic convergence of European drug makers.

I was negotiating a collaboration with Bayer that would permit Bayer to conduct biological tests on the chemical library from my own company, Cyanamid. Basically, that meant the Germans would be permitted to use our vast collection of molecules in their own drug hunting projects. During my visit, my hosts took me on a tour of Bayer's archives. I held the original handwritten notebooks of August Kekulé, one of the most famous chemists in history, best known for discovering the hexagonal structure of benzene. After winding up our meetings, my chauffeur drove me back to my hotel on the outskirts of Frankfurt. He took the autobahn. While the driver was rocketing along at speeds approaching 130 miles per hour, I did my best to put nagging questions of air bag reliability out of my mind. I noticed that our route was following the Rhine, and in the hope of stifling my fear with conversation I asked my host how Europe's most enduring drug firms all came to be concentrated along a single river. The reason, my German colleague informed me, has everything to do with the invention of colors like naphthol yellow, croceine orange, and methyl violet.

For thousands of years, humans colored their fabrics using dyes made from plants and animals. The most vibrant colors, like Tyrian purple (made from predatory sea snails) and crimson (made from scale insects), could be so costly that fabrics stained with these shades often became status symbols reserved for aristocrats and royalty. But in the early nineteenth century, the British scientist John Dalton proposed the theory of the atom, which held that there was a set of indivisible chemical elements that combined with each other according to strict mathematical laws. Dalton's atomic theory galvanized the fast-growing field of chemistry by providing a rational framework for understanding the individual components of any given chemical. After Dalton, scientists realized that every compound was composed of a specific set of molecules.

Using this new way of thinking, drug hunters could finally unravel the key constituents of many ancient medications and determine the precise purity of any given formulation. Before scientific chemistry, the true substance of flowers, trees, and plants was both unfathomable and indistinguishable. Many scientists speculated that there was some kind of mystical élan vital—the force of life—that infused plants with a kind of botanical soul. There were no principles that could explain why one particular flower was poison and another palliative. Though apothecaries had always possessed many recipes for preparing drugs from plants, they usually lacked any understanding of what the active agent actually was in any given preparation. But once it was grounded in atomic theory, chemistry finally provided a set of practical tools for determining which molecules a drug was actually made of—and which of these molecules were active. Soon chemistry was able to do even better.

By the 1830s, a new subdiscipline of chemistry had emerged known as synthetic chemistry. Synthetic chemists were able to combine simple chemical elements together into more complex compounds, like fastening together increasingly elaborate Tinker Toys. And the first businesses to harness synthetic chemistry for big profits were the dye companies.

In 1856, an English teenager named William Henry Perkin, the son of a carpenter, was experimenting with the nascent techniques of synthetic chemistry in his small apartment—not much different than a high school student playing with a home chemistry kit today. While attempting to synthesize quinine, he serendipitously noticed that one of the resulting chemicals was a bright purplish color that readily dyed silk. He dubbed the never-before-seen color "aniline purple," but the French would eventually rename it mauve. It was the world's first synthetic dye. In a matter of years, mauve launched a massive international synthetic dye industry.

For the first time, instead of relying on expensive plants or animals to produce natural dyes, companies could create fabric dyes by mixing chemicals in a lab. Even better, dye companies quickly discovered that by tweaking the chemical formula of one color they could easily get another color, offering up a seemingly boundless kaleidoscope of unimagined hues. Simply adding a couple atoms onto a molecule of red dye produced spectacular new shades of indigo, crimson, or violet. Since synthetic dyes could be manufactured in factories using highly efficient and scalable processes, they cost significantly less than traditional vegetable dyes. Fashion was transformed forever. For the first time, the middle class and even people of low means could afford garments bedecked with vivid, attractive colors. Everyone could dress like royalty.

Although the first synthetic dye was discovered in London by Perkin, nineteenth-century Germany possessed a muscular capitalist culture and a sophisticated scientific community, including many of the world's top researchers and institutions in the rapidly growing field of chemistry. As a result, the German dye industry quickly rose to global prominence as purveyors of quality synthetic dyes. (By 1913, Germany was exporting 135,000 tons of dye; Britain exported 5,000 tons.) And, finally, we can return to the Rhine. Most of the German dye factories sprang up along the Rhine because of its proximity to major European cities and because the river allowed for easy transport of both raw supplies and completed products across Germany, central Europe, northern Europe, and the rest of the world through the river's egress into the North Sea.

The Rhine dye companies not only became the world leaders of synthetic dye production, they also became the undisputed masters of synthetic chemistry, their cutting-edge research funded by profits from a public hungry for color. One of the most successful of these companies was Friedrich Bayer and Company. By the early 1880s they were selling hundreds of dyes to fabric manufacturers, but their executives had begun to look for new types of products that could leverage their growing expertise in synthetic chemistry. One of these Bayer executives, Carl Duisberg, set his sights on medicine.

Duisberg joined Friedrich Bayer and Company in 1883 with a doctorate in chemistry. As part of his military service in Munich, he had previously worked in the laboratory of Adolf von Baeyer, a famous German chemist who later won the Nobel Prize for synthesizing the color indigo (Baeyer was unrelated to Bayer company founder Friedrich Bayer.) Duisberg had been hired by the Management Board Chairman of Bayer, who was searching for young gifted chemists who could "make inventions" using synthetic chemistry that Bayer could convert into lucrative products. In 1888, Duisberg established the Bayer Pharmaceutical Research Group with a mandate to invent new medicines.

For centuries, all drug hunters—including physician-botanists, physician-alchemists, and industrial formulators—took it as a given that drugs could only be *discovered,* like a vein of gold or a hot spring, rather than *crafted* through human ingenuity, like a steam engine or a typewriter. The notion that it might be possible to *engineer* a drug to fight a particular malady required an enormous shift in perspective, and the first step in this shift was propelled by the newfound power and precision of synthetic chemistry.

Up until this point, all the industrial pharmaceutical companies (such as Squibb) were focused on using chemistry to manufacture known drugs more efficiently and consistently. But Duisberg didn't just want to improve the manufacture of existing drugs—he wanted to create drugs that had never existed before. The basic model of the synthetic dye business was to start with some molecules known to produce pretty colors and chemically tweak them to make even prettier colors. Duisberg asked, Why not do the same thing with medicines? Start with a good drug and chemically tweak it until it became an even better drug. One of Bayer's first candidates for this speculative tweaking was a commonplace drug known as salicylic acid.

Salicylates had been used for thousands of years to reduce fever, pain, and inflammation, and, like most drugs up that point, they were extracted from the library of botany. Salicylic acid was derived from vascular plants, large plants and trees such as the willow tree with nutrient-conduction systems that function like the circulatory system in animals. (Ironically, the fact that an extract from the willow tree cured fever seemed to fulfill a common principle in medieval drug hunting known as homeopathy. According to homeopathy, the cure for any given disease was found in the same place where the disease was contracted. For example, since swamps often produced fever, it was believed that the cure for these fevers would also be found in swamps. Since the willow tree was native to marshy terrain, the fact that a willow extract cured fever made a kind of sense to many eighteenth-century apothecaries.) The key ingredient from these vascular plant extracts remained unknown until 1838, when the Italian chemist Raffaele Piria developed a method for obtaining a more potent form of the willow extract, which he named salicylic acid after the Latin word for the willow tree, *salix*. Another chemist soon discovered that the active component in extracts from the meadowsweet, another vascular plant, was the same salicylic acid identified by Piria.

As physicians became increasingly aware of the benefits of salicylate medicines and improved the effectiveness of dosing, the use of these drugs accelerated through the middle of the nineteenth century until salicylics became a standard component of every physician's medicine bag. Even so, salicylic acid produced highly unpleasant side effects, particularly gastric irritation, tinnitus, and nausea. If Duisberg could find a way to reduce the side effects of salicylic acid while retaining its anti-inflammatory properties, then Bayer would have a chance to improve the drug—and make a fortune. All it needed, Duisberg hoped, was the right chemical tweak.

Even in this earliest of incarnations, the rudimentary Bayer drug development group was quite similar to the drug development teams found in Big Pharma today. There was a *chemistry team* composed of chemists who synthesized the compounds, and a *pharmacology team* composed of biologists who tested the compounds in animals and—if the animal tests were promising—in humans. Duisberg hired two lieutenants to run the salicylic tweaking efforts, Arthur Eichengrün to head the chemistry research and Heinrich Dreser to head the pharmacology research.

In general, organic compounds produced by plants are extremely complex and difficult to manipulate in a laboratory. It was Duisberg's good fortune, however, that the salicylates were unusually good candidates for tweaking, for they are rather simple molecules that are easier to manipulate than most plant compounds. In the mid-1890s, Eichengrün, the head chemist, became interested in acetyl groups, small molecules with two carbon atoms that could be attached to many plant compounds, including the salicylics. In August of 1897, Eichengrün instructed Felix Hoffman, a junior chemist in his department, to add acetyl groups to two prominent plant-derived drugs, morphine and salicylic acid. Hoffman added an acetyl group to morphine (derived from poppy flowers) and created a new synthetic compound called diacetylmorphine. He also added an acetyl group to salicylic acid (derived from meadowsweet) and created a new synthetic compound called acetylsalicylic acid.

These two new drug candidates, diacetylmorphine and acetylsalicylic acid, were sent to Dreser (the head pharmacologist) for evaluation on animals and humans. Both synthetic compounds passed Dreser's initial animal trials. But Dreser feared that he did not have a large enough budget to pursue a complete evaluation for both compounds. He believed that, given his limited resources, he could pick only one drug to develop. But which one?

My first boss in the pharma industry taught me that the most difficult and important decision in drug hunting is the decision to "fish or cut bait"—whether to continue investing resources in the pursuit of a potential drug or cut your losses and move on. These decisions are always based on inadequate information, so scientists often end up chasing after bad drugs instead of good, commercial ones. The frequency of the wrong decision to continue fishing helps explain why 50 to 75 percent of all clinical trials fail.

On the other hand, the mistaken choice to "cut bait" occurs even more frequently. When I was at Squibb I was trying to develop an alternate version of an existing antibiotic that was effective but somewhat toxic. I believed our initial work showed significant promise, but research management overruled me and shut us down before we could start our clinical trials. They decided to cut bait. Our competitor, Lilly, was trying to develop a similar antibiotic, but unlike Squibb, they decided to keep fishing. Their antibiotic eventually received FDA approval and is currently generating over $1 billion in annual sales.

Back to Dreser, head of pharmacological research at Bayer. With regard to diacetylmorphine and acetylsalicylic acid, he felt he needed to keep fishing with one of them and cut bait with the other. Dreser was more concerned about spending resources on acetylsalicylic acid because of salicylic acid's reputation for weakening the heart, which he feared would remain a side effect in the tweaked version. He judged that the morphine tweak was the more promising candidate and redirected all of his efforts into the development of diacetylmorphine, which Dreser renamed "Heroin."

Eichengrün (Bayer's head chemist) arrived at the opposite judgment. He felt that if there were only resources to pursue a single compound they should keep fishing with acetylsalicylic acid, since there would be almost unlimited applications for an effective remedy that reduced fever and relieved pain. He did not have any hard evidence showing that the salicylic acid tweak would not produce side effects, however. In order to demonstrate that acetylsalicylic acid was safe and effective, he needed data from human trials—and Dreser was blocking any further clinical trials on the drug. Eichengrün knew he could appeal to their shared boss, Duisberg, but Eichengrün also knew that Duisberg held Dreser in high esteem. Not only that, in the team-focused culture of German business, it would have been highly unlikely that Duisberg would override the judgment of a man he had just put in charge of Bayer's biological research. Even today, German drug companies abhor loose cannons and lone wolves. Eichengrün felt the pressure to toe the company line, but since he was convinced that the commercial potential of acetylsalicylic acid was simply too great to ignore he did something that daring drug hunters have always done—he went behind management's back.

Eichengrün approached a friend and colleague named Felix Goldmann, Bayer's representative in Berlin, and quietly arranged for low-profile human trials of acetylsalicylic acid in Germany's capital. This was truly the very dawn of human drug trials, so modern ethical concepts like informed consent had not yet been conceived, let alone implemented. Berlin doctors (and dentists) simply took the unidentified compound that Goldmann handed them and fed it to their patients. One dentist tested Eichengrün's compound on a patient with a toothache and reported that a few minutes later, "He jumped up saying the toothache was completely gone." Since fast-acting anti-inflammatory drugs did not exist in any form, both Eichengrün and the dentist regarded the patient's speedy relief as near-miraculous. Further tests of acetylsalicylic acid on other patients were also highly encouraging: subjects reported relief from pain, fever, and inflammation, and—crucially—they did not report gastrointestinal distress or other notable side effects.

Eichengrün shared his surreptitiously obtained findings with Dreser. Dreser was unimpressed. After reading Eichengrün's clinical reports on acetylsalicylic acid, Dreser wrote, "This is the usual loud-mouthing of Berlin—the product has no value." He firmly believed that Heroin represented the future of the company. Duisberg finally intervened in the dispute between his two top lieutenants. He reviewed Eichengrün's Berlin data, overturned Dreser, and authorized the full clinical testing of acetylsalicylic acid on humans—alongside the full clinical testing of Heroin.

Both synthetic drugs passed human trials with flying colors and Bayer prepared to sell them to the public. In 1899, Bayer selected the commercial name for acetylsalicylic acid: Aspirin. The name is derived from the *a* in acetyl, plus the Latin name for the meadowsweet plant (*Spirea ulmaria*), and the standard drug suffix *-in*, which was believed to make the name easier to pronounce across European languages. Bayer also made sure that the generic term for Aspirin was as difficult to pronounce as possible: monoacetic acid ester of salicylic acid.

But Bayer ran into an unexpected and discouraging wrinkle. Since other researchers had previously reported the synthesis of acetylsalicylic acid, Bayer's application for a patent in Germany was rejected. Just as Robert Talbor tried to suppress competition from other chinchona bark peddlers in the 1600s by claiming that his Pyretologia drug contained secret ingredients, Bayer pushed its convoluted name for the generic version of Aspirin to help discourage physicians from prescribing the generic version instead of Bayer's brand—the company hoped doctors would be reluctant to instruct their patients, "Take two monoacetic acid esters of salicylic acid and call me in the morning."

Even though it did not have an exclusive patent on the drug in Germany (Bayer did get one in the United States), it launched a very heavy marketing push, and Aspirin soon became the first blockbuster drug of the Age of Synthetic Chemistry. It was far superior to the old salicylate drugs derived from plant extracts. Aspirin performed just as well but exhibited markedly reduced side effects. Its global popularity grew still further when it became a standard treatment during the Spanish flu pandemic of 1918. After Bayer's American patent on Aspirin expired in 1917, there was an explosion of aspirin generics and knock-offs, but as you know from any visit to your local CVS or Walgreens, Bayer's formulation of Aspirin has remained a steady seller, one of a small handful of nineteenth-century drugs that have survived unchanged into the twenty-first century.

Today, more than 70 million pounds of aspirin are sold each year, about the weight of a small aircraft carrier. The use of the drug has slowly decreased over time because of competition from several other over-the-counter analgesics, especially Tylenol, Advil, and Motrin. Aspirin remains unique among its competitors, however, because it also thins the blood by reducing the aggregation of platelets, so it still maintains enviable sales as a heart medication.

Today, if you turn to any account of the origin of Aspirin in a contemporary textbook or history of medicine, you will almost always find that Eichengrün's name is curiously absent, even though he was single-handedly responsible for pushing Bayer to make the drug. Instead, Eichengrün's junior chemist Felix Hoffman is typically celebrated as the drug's inventor. According to the standard narrative, Hoffman developed Aspirin to help his father, who was suffering from the side effects of the sodium salicylate he took for his rheumatism. In truth, Hoffman was a very minor character in the story of Aspirin, someone who simply obeyed Eichengrün's request to add an acetyl group to salicylic acid without even knowing exactly why he was synthesizing the compound. So how did the popular account come to vary so widely from the truth? You can blame the Nazis.

Bayer did not publish the story of the discovery of Aspirin

until the early 1930s. This delay was due in large part to Bayer's head biologist, Dreser. He never forgave Bayer's head chemist, Eichengrün, for going behind his back to test Aspirin, and when he reported the company's scientific findings to help publicize the new drug, Dreser spitefully omitted any mention of Eichengrün at all. When Bayer finally published its public account of the discovery of Aspirin nearly fifty years after Eichengrün successfully guided it through, the headache remedy had become something of a national treasure. Unfortunately for Eichengrün, the Nazis had gained power in Germany, which meant that national treasures had to conform to Aryan ideals.

Even though Eichengrün by this time had become a prominent industrialist running his own chemical company, he was also a Jew. He was eventually interned at the concentration camp in Theresienstadt, where he languished until his liberation by the Soviet Army. When Bayer published its official story of Aspirin's discovery, the company prudently overlooked the fact that a Jew had been the driving force behind the drug and instead assigned credit to Hoffman, an acceptably Aryan German. During the Nazi era, the Hall of Honor in the chemistry section of the German Museum in Munich featured a showcase filled with white crystals and emblazoned with the declaration, "Aspirin: inventors Dreser and Hoffmann."

After the war, the octogenarian Eichengrün published several accounts of the true story, supporting his version with original documents. For his part, Hoffman never publicly took credit for Aspirin and never disputed Eichengrün's accounts. Nevertheless, the Nazi-influenced story of Aspirin's discovery had become too entrenched within the history of chemistry, and Eichengrün's efforts at setting the record straight were mostly ignored.

In many ways, the false history of Aspirin is an apt metaphor for the gap between the public's perception of how drugs are discovered and the far-grittier reality. In the sanitized version, Felix Hoffman invented a new drug to help his ailing father, and his brilliant discovery was quickly recognized by Bayer, who immediately shared it with the world. In reality, a vindictive middle-manager preferred the commercial prospects of Heroin over Aspirin, and he did everything he could to shut Aspirin down. Meanwhile, Aspirin's inventor came up with a clandestine scheme to obtain data on the drug (a scheme that would be considered highly unethical by today's standards) in order to go over his colleague's head and persuade senior management to back Aspirin. Then, even after Aspirin was launched, it turned out the drug was not actually a new invention at all, since Aspirin had already been synthesized by several other chemists. Despite a gush of generic competitors, Bayer still managed to squeeze blockbuster profits from its synthetic compound through savvy marketing—and then laundered the story of the drug's discovery to conform with the anti-Semitic politics of early twentieth-century Germany.

That is the real story behind the brand-name drug that has sold more units than any other in history—the drug that inaugurated the hunt through a new and untapped library of molecules, the library of synthetic medicine.

5 The Magic Bullet

We Figure Out How Drugs Actually Work

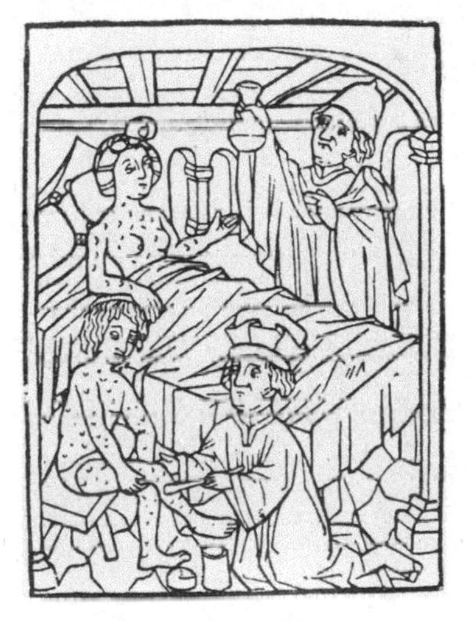

Early depiction of syphilis

"*Corpora non agunt nisi ligata.* (A substance is not effective unless it is linked to another.)"

—Paul Ehrlich, 1914

In the waning years of the fifteenth century, a new epidemic swept through Europe like a foul wind. The disease first made itself known as angry red ulcers blooming upon the skin. Rather disconcertingly, these cankers usually started on the genitals. Before long, the patient would develop a scarlet-pink rash on his chest, back, arms, and legs. A fever, accompanied by a headache and sore throat, came next. The afflicted would lose weight, then lose hair. But then, after a few weeks of steadily worsening health, the symptoms would abruptly subside. Did the body fight off the infection? No. The reprieve represented false hope.

It was not the end of the storm, but the quiet eye at the center of a biological hurricane. After a short period of time the disease surged back in horrible fashion. Hundreds of tumorous balls burst forth from the skin, red and misshapen, making the victim resemble some fairytale demon. Eventually the disease attacked the heart, nervous system, and brain, often producing total dementia. And then—sometimes after a few years, sometimes a few decades—respite usually came at last, in the form of death.

The first well-documented outbreak of the disease in Europe occurred in 1494 among French troops besieging Naples. The Italians called it "the French disease." The French, on the other hand, called it "the Italian disease." Today we call it syphilis. Since syphilis is easily confused with other diseases (it is often called the "great imitator"), its precise origins are still debated. One prominent theory holds that when Columbus and other early European explorers introduced the scourge of smallpox to the aboriginal peoples of the New World, they simultaneously carried syphilis back to Europe; the Italian outbreak occurred shortly after Columbus returned from his first voyage. What is known for certain is that syphilis was one of the most feared and infectious diseases in Europe from the 1500s through the early twentieth century.

The Spanish physician Ruy Diaz de Isla wrote in 1539 that more than a million Europeans were infected with the ghastly syndrome. Treatment options ranged from poor to useless, such as gum of the guaiacum tree (useless), wild pansy (useless), and mercury, the best of the bad. Mercury had some ameliorative effect on the disease because of its toxicity to the syphilis pathogen. Unfortunately, mercury is quite toxic to humans, too. Nevertheless, since the compound was the only meaningful therapy for the disease, its use fostered the saying, "A night in the arms of Venus leads to a lifetime on Mercury."

When syphilis began ravaging Europe, nobody knew how to treat it, because nobody had the faintest idea what caused it—or any disease, for that matter. Until the mid-nineteenth century, the leading hypothesis about the origins of common scourges like typhoid fever, cholera, bubonic plague, and syphilis was known as *miasma theory*. Miasma theory held that disease was caused by a noxious form of "bad air." This pestilent miasma presumably emanated from decomposing organic matter as a toxic mist filled with rotten particles. *People* were not infectious, according to the theory; rather, disease emanated from a *place* that gave rise to infectious vapors, identifiable by their putrid aroma. Since hospitals, by definition, were clean places that lacked any source of miasma, hospitalized patients were believed to be free from the risk of new infection.

The miasma theory was challenged in 1847 by Ignaz Semmelweis, a Hungarian obstetrician who worked at the Vienna General Hospital. He frequently treated women with puerperal fever, also called childbed fever. The disease often developed into puerperal sepsis, a serious blood infection that was sometimes fatal. We now know that this disease is caused by a bacterial infection contracted by women during childbirth, but in the nineteenth century doctors were baffled by its persistent presence in maternity wards.

Semmelweis wondered why so many new mothers were getting sick. He noticed that many women who delivered their babies at the hospital with the aid of doctors and medical students soon died of puerperal fever. On the other hand, there were no deaths among women who gave birth when they were solely attended by midwives. This was a strange enigma that defied easy explanation, but Semmelweis offered up a bold hypothesis.

He noticed that physicians and medical students often came to the obstetric wards directly after conducting an autopsy. He speculated that there was some kind of contagion present in the autopsy material that was transmitting puerperal fever to the women. To test this radical theory of direct physical contamination, Semmelweis ordered the doctors in his maternity ward to scrub their hands with lime before examining pregnant women.

No longer were physicians handling dead flesh, then moments later touching women's private parts with unwashed hands. It was a success. After Semmelweis's experiment, childbirth mortality plummeted from 18 percent to 2 percent.

Semmelweis's improvements in physician hygiene seemed to disprove miasma theory, pointing to a new way to think about disease. Unfortunately, Semmelweis and his theories were soundly rejected by the Viennese medical establishment. In 1861, Semmelweis published a book to defend his views, *Die Ätiologie der Begriff und die Prophylaxis des Kindbettfiebers* (The Etiology, Concept, and Prophylaxis of Childbed Fever). The book was mostly ignored, though it was occasionally ridiculed by more distinguished physicians who believed that Semmelweis was a bush-league dilettante.

The professional humiliation that Semmelweis endured reminds me of something that happened at a prestigious academic biology conference I attended on Long Island. The conference mostly focused on DNA, and one young postdoc gave a talk about how the extremely long strands of human DNA (they are almost ten feet long, despite being only two nanometers wide) could be squeezed into the tiny space of a microscopic cell nucleus. The young man lacked confidence and his presentation was uneven, but his findings, we know today, were essentially correct.

Suddenly, in the middle of the postdoc's talk, Francis Crick walked to the front of the stage. Crick was one of the discoverers of the structure of DNA and is one of the most famous biologists in the world. Crick stood directly in front of the lecture podium facing the young man. Their noses were only about a foot apart. Despite becoming unnerved by this bizarre display by one of the legends of science, the postdoc somehow managed to hurry through the rest his talk. As soon as he stopped speaking, Crick spoke up.

"Are you quite finished?"

The young man nodded. Crick slowly turned to face the audience and declared, "I do not know about anyone else, but this is about all the amateurism I am willing to stomach at this meeting." I imagine that Semmelweis must have felt just as humiliated as that aspiring young biologist.

Frustrated by his colleagues' dismissal of his ideas, Semmelweis began to decry obstetricians as thoughtless murderers. They shrugged him off. Doctors continued to insert their fingers into decomposing corpses and then casually use the same digits to deliver babies. Semmelweis began to drink heavily and soon became an embarrassment to his hospital and his family. In 1865, he was duped into entering an insane asylum, where he was locked up. He tried to escape but was severely beaten by guards. Two weeks later, he perished from his wounds. Such was the tragic life of the man who discovered the role of germs in creating infection.

Though several people over the centuries had proposed some version of the notion that disease was caused by direct physical contamination, clear and conclusive proof of the existence of contagious pathogens finally arrived in the 1860s through the work of the famed French biologist Louis Pasteur. Pasteur conducted experiments to disprove miasma theory and also to disprove spontaneous generation, the widely held notion that new life could erupt forth from inert matter. Imagine, for instance, you were gazing into your mobile device when tiny creatures suddenly came writhing out of your screen—nineteenth-century biologists believed this kind of occurrence was possible due to spontaneous generation.

Pasteur demonstrated that the generation of new life required exposure to specific types of particles in the air—and, crucially, he showed that these peculiar particles were already alive. Disease, in other words, was caused by organisms too tiny to see—*micro*-organisms. Scientists had known about the existence of microorganisms since the 1600s, but the nineteenth-century medical establishment could not imagine how something so tiny and insignificant could possibly sicken—let alone extinguish—a healthy human being.

Once Pasteur revealed that vanishingly small organisms caused some of the most terrible diseases known to humankind, everyone wanted a look at them. Unluckily for would-be germ-gazers, the cells of infectious bacteria and fungi (not to mention the cells of animals and plants) are largely translucent. If you stick a cell on a slide and peer at it through a microscope, you will see vague and indistinct contours that are difficult to resolve. The reason is that there is no *contrast*—no way to sharply distinguish the structures of the cell from its background.

A solution arrived in the mid-nineteenth century with the invention of synthetic dyes. Dye manufacturers were like the aerospace industry of the nineteenth century, producing a variety of useful spinoff products as they developed the high-tech products for their core market. Microbiologists began to test off-the-shelf fabric dyes to see if they might also be useful for staining cells. One man who was captivated by synthetic dyes' potential for improving the study of germs was a German scientist by the name of Paul Ehrlich.

Ehrlich's cousin Karl Weigert was a prominent cell biologist and histologist (someone who studies the structure of living tissue). Between 1874 and 1898, Weigert published a series of papers on the use of synthetic dyes to stain bacteria. (Even today, scientists still use the "Weigert stain" to view neurons.) Weigert's work led to the rapid adoption of a set of synthetic dyes for studying animal cells and microorganisms known as "aniline dyes." These dyes were based upon the aniline molecule, an organic compound that smells like rotten fish.

Ehrlich followed in his cousin's footsteps and began using aniline dyes to stain animal tissues in medical school in Leipzig. He obtained his medical degree in 1878, but he was never considered a particularly promising student. His professors believed his obsession with tissue staining was a pointless distraction that was preventing him from developing more useful skills. When one of Ehrlich's professors introduced Ehrlich to Robert Koch, an eminent physician regarded as the father of bacteriology for his pioneering research on infectious diseases, the professor told Koch, "That is little Ehrlich. He is very good at staining, but he will never pass his examinations." In fact, there was nothing in Ehrlich's early career to suggest that he would eventually become involved in drug hunting, let alone become one of the most influential drug hunters of all time.

Early on, Ehrlich became fascinated by the peculiar fact that some dyes would stain parts of a certain type of cell (such as the cell wall or chloroplast components in plant cells), while failing to stain any part of other types of cells (such as animal cells). In other words, each dye seemed to have its own biological target that it would adhere to. One day he was struck by a provocative idea: What if a dye that targeted parts of a certain kind of pathogen was also toxic to that pathogen? If so, it might be possible to kill the pathogen without harming the host. Ehrlich called this notion of pathogen-targeting toxins *Zauberkugeln*—"magic bullets."

In 1891, Ehrlich commenced his search for a dye that would selectively target the protozoan that causes malaria . . . and kill it. After testing dozens of dyes, he observed that one particular dye known as methylene blue stained the parasite but did not stain human tissue. Even more promising, the dye appeared to have some toxicity to the malaria pathogen. He began testing methylene blue on several malaria patients and soon reported that he had cured two of them. The world's first fully engineered drug was a bright, vibrant dye the color of cobalt.

Ehrlich admitted that quinine remained a much more effective and reliable treatment for malaria, but he had proven that his notion of a magic bullet was not mere theory—it actually worked in practice. All that was needed was the right kind of dye. He was given an appointment at the Institute for Infectious Diseases in Berlin, where he set up one of the first successful models of a drug research lab. His lab included an organic chemist who developed new drug candidates (that is, new synthetic dyes), a microbiologist who tested the effects of drug candidates on pathogens (this was Ehrlich's role), and an animal biologist who tested the effects of drug candidates on animals and—if the animal tests were successful—on humans.

Ehrlich's three-unit team studied the staining and toxicity of hundreds of synthetic dyes on pathogenic protozoa, infectious single-cell microorganisms that are more similar to mammal cells than bacteria. Though they found many dyes that selectively targeted germs, none of them impaired the protozoa's activity until they stumbled upon trypan red. This dye stained a parasite in mice known as *Trypanosoma equinum*—and killed it. Ehrlich's initial rush of excitement was short-lived, however. The trypanosome pathogens rapidly developed resistance to trypan red, rendering it useless as a cure.

After a seemingly interminable string of failures, Ehrlich realized he might need to modify his magic bullet theory. Perhaps it was simply too difficult to find a double-duty dye that both targeted a pathogen *and* slayed it. Instead, why not take a toxin that was already known to kill a pathogen, and use chemical synthesis to mount the toxin onto a dye known to target the pathogen, producing a kind of "toxic warhead"? Even if the toxin was harmful to humans, by attaching it to a dye that targeted a particular germ it could act like a guided missile, delivering its destructive payload directly to the germ.

Ehrlich launched his new toxic warhead approach to drug hunting by using arsenic as the payload. A French scientist named Antoine Béchamp had previously shown it was possible to attach a molecule of arsenic to a molecule of dye, creating a new compound called atoxyl. Atoxyl was highly toxic to humans, but Ehrlich wondered if he could synthesize a variant of atoxyl that would be safe in humans but lethal to germs. Ehrlich knew that atoxyl stained *Trypanosoma cruzi,* a parasite causing a nervous system disease known as trypanosomiasis, so for his first round of arsenic experiments Ehrlich chose to target *T. cruzi.* His team began creating hundreds of variants of atoxyl and testing them on mice infected with the parasite, but all of these synthetic warheads either failed to kill the trypanosomiasis, or if they did succeed, they killed the host, too.

Frustrated, Ehrlich switched diseases. In 1905, a zoologist working with a dermatologist identified the pathogen that causes syphilis, a spirochete bacterium known as *Treponema pallidum.* Ehrlich believed there was a biological similarity between spirochetes and trypanosomes, even though we now know that there is virtually no structural or genetic similarity at all. Nevertheless, motivated by this erroneous assumption, Ehrlich applied his atoxyl toxic warheads to syphilis.

His team synthesized more than nine hundred arsenic-loaded dyes and tested them on rabbits infected with syphilis. Each compound was a failure. As the team began to think about switching tactics once again, in 1907, Ehrlich's animal biologist noticed that one of the compounds seemed to be killing the bacterium without killing its hosts. The compound was labeled "606" because it was the sixth compound in the sixth test group. When Ehrlich reported on the success of 606 in the *New England Journal of Medicine* in 1911, he named it "arsphenamine." Clinical studies soon demonstrated that arsphenamine was an effective and safe treatment for syphilis in humans. It was, at last, a bona fide magic bullet.

Ehrlich teamed up with the German company Hoechst AG—a company that had provided him with many dyes over the years—to manufacture arsphenamine for commercial use. It was marketed to the public in 1910 under the trade name Salvarsan with the tagline, "The arsenic that saves."

Ehrlich's toxic warhead was the first reliable and effective treatment for a contagious disease. Put simply, it was the world's very first cure. But that was not the only reason that the discovery of Salvarsan represented an extraordinary moment in the history of medicine, and indeed, in the history of humankind. Never before had someone thought up a novel way to create a completely unprecedented kind of drug—and then gone out and actually made it. Salvarsan was not a better-engineered knockoff of an existing drug, like Squibb's ether, or a minor tweak of an existing drug, like Aspirin. Instead, it was the product of a wholly original conception: find a dye that stains a pathogen, then find a pathogen-killing toxin that will attach to the dye.

Almost overnight, Salvarsan became both famous and notorious. It actually *eliminated* a disease once and for all, instead of merely mitigating its symptoms. At the same time, since the disease in question was a sexually transmitted infection associated with promiscuity and prostitutes, the number *606* quickly became the butt of innumerable off-color jokes, similar to the number *69* today. Many telephone exchanges even dropped the 606 code because of its newfound sexual connotations.

Isak Dinesen, the author of the memoir *Out of Africa*, was one of the first people to be treated with Salvarsan. A Danish aristocrat whose real name was Baroness Karen von Blixen-Finecke, Dinesen managed a coffee plantation in Kenya for most of her adult life. According to her memoir, her husband was a serial philanderer who infected her with syphilis. After realizing she had contracted the embarrassing and deadly disease, Dinesen returned to Denmark for many long months of treatment with Salvarsan. Although her physicians eventually declared that she was cured, she remained dubious, probably due in part to the fact that no disease—including syphilis—had ever been curable before. Though extensive tests failed to reveal any evidence of syphilis remaining in her system, for the rest of her life she remained convinced that she was still afflicted. Even so, her exquisite writing suggests that she suffered neither from the mental degeneration characteristic of advanced syphilis nor from cerebral damage due to excessive Salvarsan treatments. Ehrlich's magic bullet enabled Dinesen to become one of the finer literary voices of the twentieth century.

The enormous success of his drug made Ehrlich a public hero. Whenever he was congratulated for his achievement, however, he modestly replied: "For seven years of misfortune I had one moment of good luck." If he had correctly understood that the syphilis pathogen and the *Trypanosoma* pathogen were vastly different microorganisms, he probably would have never tried his toxic warhead on the Italian disease. The German-born Ehrlich concluded from his experience that a drug hunter needed what he called "the four G's": *Geld* (money), *Geduld* (patience), *Geschick* (ingenuity), and perhaps most importantly, *Glück* (luck). His formula was extremely prescient, since money, patience, ingenuity, and a heaping dose of serendipity remain essential ingredients for drug discovery to this very day.

Ehrlich's method for developing Salvarsan established a completely new vision of what a drug actually was, a conception so strange and radical that the scientific community initially rejected it. Between 1895 and 1930 there were four competing theories of how drugs worked: the "physical theory," the "physicochemical theory," the "Arndt–Schulz Law," and the "Weber–Fechner Law." All four of these theories were utterly wrong. The physical theory held that the surface tension of the cells in a given tissue dictated what kinds of drugs affected that tissue. The physicochemical theory was a variant of physical theory, arguing that drugs worked by altering the surface tension of cells. The Arndt–Schulz Law postulated that drugs affected the body according to the following formulation: "weak stimuli excite, medium stimuli partially inhibit, and strong stimuli produce complete inhibition." Needless to say, this wooly-headed hypothesis had little connection with biochemical reality. Finally, the Weber–Fechner Law hypothesized a logarithmic relationship between the size of a drug's dose and the size of its effect, an idea rather incongruously drawn from a theory of human perception. None of these theories was remotely accurate, and worse, none of them provided any guidance on how drugs could be improved or how new drugs might be discovered.

But Ehrlich developed a new way of thinking about drugs that he succinctly summarized in the Latin phrase *Corpora non agunt nisi ligata*—"a substance is not effective unless it is linked to another." He called this new conceptual framework "side-chain theory," a conception that grew out of Ehrlich's understanding of the human immune system. He correctly hypothesized that a person's immunity to a disease was based on the reaction of special substances in a person's blood serum to the toxic compounds in a pathogen. He called these substances "side-chains," though today we call them "antibodies," and we call the toxic compounds that the antibodies react to "antigens."

Ehrlich argued that a particular antibody binds to a particular toxin in a lock-and-key fashion and that this selective chemical binding triggered the immune system to eliminate the pathogen, a theory that we now know to be accurate. He extended the same lock-and-key thinking to drugs, believing that there was a specific molecular site on a pathogen or on a human cell (the "receptor") that reacted with a specific part of a drug, thereby producing its effect. This is known today as "receptor theory."

Ehrlich's novel conception of drug action was based on his discovery that chemical dyes only stained particular parts of cells, and his receptor theory now serves as the foundation for modern pharmacology. But in 1897, when Ehrlich first proposed receptor theory, he could not provide any direct evidence for the existence of receptors, which he claimed were too small to be visible under existing microscopes. Not surprisingly, other scientists regarded his idea of invisible antibody receptors as squatting somewhere between pseudoscientific and preposterous.

A group of scientists at the prestigious Pasteur Institute in Paris helped lead the opposition against receptor theory. For ten years, the Pasteur scientists conducted experiments on blood proteins that they contended disproved receptor theory. Ehrlich performed the exact same experiments and obtained similar results, but argued that they actually validated his theory. Since the details of these experiments were very complicated and involved sophisticated scientific reasoning, most scientists simply tended to believe the more straightforward arguments coming from the highly regarded Pasteur Institute.

Feeling increasingly aggrieved, Ehrlich became an obsessive

and bristly defender of his ideas, sorting all of his colleagues into "friends" or "enemies" depending on their view of receptor theory. In 1902, for instance, he wrote to William Henry Welch: "I was most delighted to recognize you as one of the warmest friends of the theory, but even more that you could achieve such new and fundamental insights with its help." In contrast, he wrote to a pharmacologist in Halle, Germany, "that every impartial person reading the literature has to count you as an absolute opponent."

One of the most formidable opponents of receptor theory was Max von Gruber, a famous professor of hygiene at the University of Munich. Nobody else possessed the same ability to infuriate Ehrlich. While Gruber acknowledged Ehrlrich's contributions to the nascent field of immunology, he published several papers that attacked Ehrlich's receptor theory of drugs as purely speculative, burdened by "a nearly total lack of evidence." Gruber's concerns were rather reasonable given the inability of scientists at the time to identify any drug receptors in the human body. Nevertheless, Ehrlich castigated the hygienist's criticisms as "stupid" and "negligible." On one occasion, Ehrlich was evicted from a train because of his loud complaints about Gruber. The more level-headed Gruber responded by writing, "I only reproach Ehrlich for permitting too much fantasy in his theories while accepting too little criticism."

Though Ehrlich's theory was ultimately proven correct, it took almost a century before the underlying details of receptor theory were fully understood. When I first studied pharmacology in the 1970s, the definition of a receptor was still tautological: the "adrenaline receptor" was the thing that bound to adrenaline. I had previously studied biochemistry and molecular biology, well-developed fields where scientists knew with tremendous precision the intimate details of the molecules they were manipulating. Biochemists could usually specify exactly how one compound would interact with another compound. In comparison, pharmacologists usually possessed only a shockingly vague idea of how their drugs worked. For example, the receptor that aspirin acted upon had just been identified a few years before I started my pharmacology studies, more than seventy years after aspirin was first used to treat patients.

We now know that most receptors within the human body are protein-based molecular switches that turn cellular processes on or off by reacting to hormones in the body. For example, there are a number of distinct adrenaline receptors in the human body, including the beta-2 receptor, a protein present in smooth muscle cells that reacts with adrenaline to produce muscle relaxation. Once scientists identified the beta-2 receptor as an adrenaline receptor, drug hunters began to search for medications that activated them. One of the best-known drugs to come out of this search was albuterol, used as an inhaler by asthmatic patients. Albuterol opens up a person's air passages by relaxing the smooth muscle cells in the lungs, improving breathing and preventing or mitigating an asthma attack.

Though most scientists were skeptical of Ehrlich's theory of how drugs worked, there was no denying the astonishing effectiveness of Salvarsan—or Ehrlich's wholly original method of engineering Salvarsan by attaching a germ-killing compound to a molecule of dye. It was the culmination of the Age of Synthetic Chemistry, the first proven method for creating drugs from scratch rather than discovering drugs in the library of plants or by tweaking existing drugs.

You might imagine that Ehrlich's ingenious cure for syphilis ushered in a golden age of drug hunting as pharma scientists around the world engineered their own magic bullets. You might, but you would be wrong.

6 Medicine That Kills

The Tragic Birth of Drug Regulation

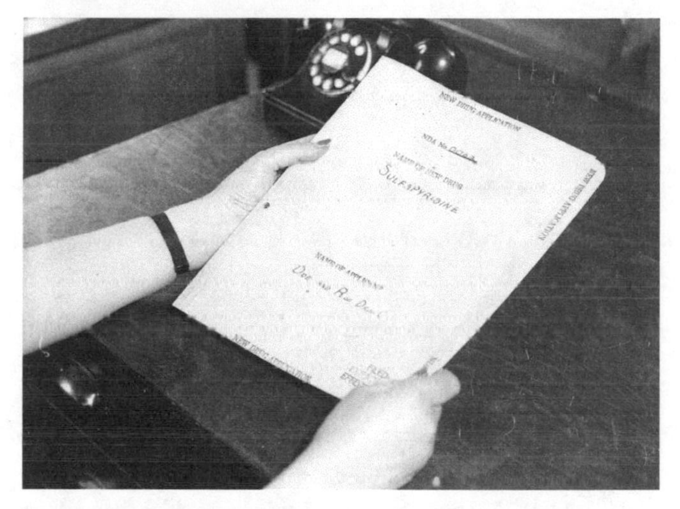

The first drug application submitted to the FDA

"We have been supplying a legitimate professional demand and not once could have foreseen the unlooked-for results. I do not feel that there was any responsibility on our part."

—Samuel Evans Massengill, 1937

Ehrlich's discovery of Salvarsan in 1909 established a rational and methodical approach to drug hunting. It showed that it was possible to design and synthesize a new drug from scratch by thoughtfully applying one's knowledge of chemistry and biology. Salvarsan also served as a drug hunting milestone in another important respect. The compound 606 was the first successful antibiotic. When Ehrlich loaded an arsenic warhead onto a khaki-colored dye, there was no reliable, effective cure for an infectious disease. Physicians could sometimes ameliorate the symptoms of various afflictions, but they possessed no confident remedy. After Ehrlich, everything changed. Salvarsan provided doctors with an unprecedented weapon that actually destroyed the *source* of a disease, the syphilis bacteria.

Even so, Salvarsan had considerable shortcomings. The drug needed to be dosed with exceptional care. If too little was administered, the syphilis bacterium would not die. If too much was administered, the patient could die. And the drug was not effective at all if the syphilis had already progressed to an advanced state. But the drug's chief limitation was the fact that it worked only on a single disease—syphilis.

Today, we enjoy the benefits of many "broad-spectrum antibiotics" such as penicillins and fluoroquinolones, drugs that fight a wide range of infectious pathogens. But Salvarsan was a "narrow-spectrum antibiotic"—a one-hit wonder. At the time of Ehrlich's great discovery, there was not yet a clear notion that it might be possible to develop drugs that attacked multiple types of infections. Instead, the focus was on discovering *any* new cure, whether it turned out to be a silver bullet or a silver shotgun blast. Inspired by Ehrlich, a new generation of drug hunters was spurred to search for other synthetic anti-infectives. The largest pharmaceutical laboratories of the early twentieth century put their top researchers to work screening for bacteria-killing dyes, especially the German companies on the Rhine. The rush for synthetic cures began with great enthusiasm, with many chemists predicting a golden age of drug discovery.

This glow of optimism gradually darkened. After twenty years of well-funded drug hunting, not a single new antibiotic had been found. By the early 1930s, it was beginning to appear as if Ehrlich had been indescribably lucky in his discovery of a synthetic Vindication. Scientists were beginning to suggest that Salvarsan was one-of-a-kind. Then, in 1935, Bayer AG—a corporate descendant of the chemical company that created Aspirin—finally struck gold. Bayer AG had assembled yet another research team in 1932 to try to crack the riddle of making a general-purpose antibiotic from an aniline dye. The team had tested out several thousands of dyes on several thousands of mice, and none showed promise. Then, one day, they tested a bright red dye. It killed several different kinds of infectious bacteria. Bayer dubbed their new drug Prontosil.

Prontosil was the first broad-spectrum antibiotic. It cured a variety of maladies caused by streptococci bacteria, including blood infections, skin infections, and childbed fever. However, there was something quite perplexing about the drug. It worked only on living animals or living people. The drug failed to kill bacteria growing in test tubes. This presented a new mystery for Bayer AG: Why did Prontosil extirpate pathogens in the body but fail to dispatch the same pathogens when they were outside of the body?

This pharmacological riddle was finally solved by a Pasteur Institute research group, who discovered that when Prontosil was metabolized in the liver, it was broken down into several smaller compounds. One of these compounds was a colorless molecule known as sulfanilamide. The Pasteur Institute scientists showed that the large Prontosil molecule itself had no effect at all on bacteria. Instead, it was the much smaller sulfanilamide molecule that was the true antibiotic. It wiped out bacteria inhabiting both living creatures and Petri dishes. The reason Prontosil failed to destroy bacteria outside of the body was that it had not been broken down into its active component.

Bayer's great triumph—the creation of the first broad-spectrum antibiotic—was based on a false premise, the idea that a toxic dye was selectively targeting bacteria, the way Salvarsan did. Instead, it turned out to be pure biochemical chance that the mammalian physiology transformed the red Prontosil dye into an entirely new compound that cured infections. While this discovery was scientifically embarrassing to Bayer, financially it was devastating. Sulfanilamide was a familiar compound that had been used by chemists for decades and thus could not be patented. The day after the Pasteur Institute published their findings about sulfanilamide in 1936, chemical manufacturers around the world woke up to discover there was a miracle drug that anyone could legally make and sell.

Within a few years, hundreds of companies were churning out their own idiosyncratic versions of sulfanilamide, launching an international "sulfa craze." One of the myriad new formulations of sulfa was Elixir Sulfanilamide, produced by the Tennessee pharmaceutical manufacturer S. E. Massengill Company. The company had been founded in Bristol, Tennessee, in 1898 by one Samuel Evans Massengill, a graduate of the University of Nashville Medical School. His firm had manufactured everything from analgesics to ointments before trying to cash in on the sulfa craze, usually marketing Massengill's products with names that were variants of his own, such as Anagill, Dermagill, Giagill, Resagill, and Salogill.

Massengill's sulfanilamide preparation was simple. Sulfanilamide was dissolved in diethylene glycol, then raspberry flavoring was swirled in. This preparation was formulated by the chief pharmacologist at S. E. Massengill, a man named Harold Watkins. Though Watkins was a trained chemist, his training had apparently not led him to the awareness that the sweet-tasting diethylene glycol was a strong poison. (Today, diethylene glycol is used in brake fluid and wallpaper stripper.)

Animal testing was already fairly widespread in the pharmaceutical industry by the 1930s, but in his rush to get Elixir Sulfanilamide to the market Watkins did not bother to test his formulation on any living creature. There was nothing illegal about this seemingly egregious oversight—there were no laws requiring *any* testing before selling a drug to the public. Though the Food and Drug Administration had been established by Congress in 1906, the agency was largely toothless. Its main purpose was to ban adulterated or mislabeled products rather than enforce safety.

Elixir Sulfanilamide went on sale in drugstores around the country in September of 1937. The Reverend James Edward Byrd of Mount Olive, Mississippi, was one of the first customers to buy a bottle. Byrd was a sixty-five-year-old Baptist preacher and the long-serving secretary of Mississippi's Baptist Sunday School. On October 11, he had consulted his good friend Dr. Archibald Calhoun about his cystitis, a painful infection of the urinary tract. Calhoun prescribed sulfanilamide, which remains a safe and highly effective treatment for cystitis. Byrd went to his local pharmacist, who fulfilled his doctor's prescription with a bottle of S. E. Massengill's Elixir Sulfanilamide. (Dr. Calhoun also prescribed the Elixir for five other people.)

After taking the prescribed dose, Byrd departed for a series of clergy meetings in Knoxville. The next day Byrd "felt a constant urge to urinate," but found it "difficult to start the stream and very little was voided." A few days later, as he continued to have difficulty passing water, Byrd was admitted to the Knoxville hospital. He was diagnosed with catastrophic kidney failure. The staff administered intravenous saline and glucose in an emergency attempt at stimulating his renal function, to no avail. The reverend's wife, Leona, and his two sons were at his side as he died an excruciating death.

Two doctors from the University of Chicago published a paper in the *Journal of the American Medical Association* that concluded that Byrd's demise had been due to diethylene glycol, a compound known to devastate kidneys. Byrd's physician, Calhoun, become despondent. He wrote a letter to President Franklin D. Roosevelt:

> Any doctor who has practiced more than a quarter of a century has seen his share of death. But to realize that six human beings, all of them my patients, one of them my best friend, are dead because they took medicine that I prescribed for them innocently, and to realize that that medicine which I had used for years in such cases suddenly had become a deadly poison in its newest and most modern form,

> as recommended by a great and reputable pharmaceutical firm in Tennessee: well, that realization has given me such days and nights of mental and spiritual agony as I did not believe a human being could undergo and survive.

More than one hundred people around the country died from S. E. Massengill's sulfa formulation, including many children who had been prescribed the elixir for sore throats. One of these children's mothers, Mrs. Maise Nidiffler of Tulsa, Oklahoma, also wrote to Roosevelt:

> The first time I ever had occasion to call in a doctor for [Joan] and she was given Elixir of Sulfanilamide. All that is left to us is the caring for her little grave . . . we can see her little body tossing to and fro and hear that little voice screaming with pain and it seems as though it would drive me insane. . . . It is my plea that you will take steps to prevent such sales of drugs that will take little lives and leave such suffering behind and such a bleak outlook on the future as I have tonight.

In the 1930s, the federal government treated drugs the same way they treated paper clips or trousers, as products that required no special safety regulations. The American Medical Association (AMA) was not involved in approving drugs either—the leading professional medical organization merely shared information they received voluntarily from drug manufacturers or physicians about particular drugs. S. E. Massengill had not shared any information about Elixir Sulfanilamide, and therefore the AMA possessed none.

After reports of patient deaths from Elixir Sulfanilamide reached the AMA, they telegraphed Samuel Evans Massengill himself to request the composition of his company's drug. Massengill admitted that it contained diethylene glycol, but insisted that this fact be kept strictly confidential—though not because he believed the solvent was dangerous. He worried that other companies might try to steal his formulation for themselves. When the AMA confronted him about the growing number of deaths from Elixir Sulfanilamide, Massengill and his chief chemist Watkins confessed that no toxicity tests had been performed but proposed that the fatalities might have resulted from taking the elixir with other drugs in some harmful combination. To demonstrate his confidence in his product, Watkins swallowed small amounts of Elixir Sulfanilamide and was "pleased to report that I have noted no adverse effects."

Less than two weeks after Watkins' blithe self-experimentation, there was an abrupt about-face at his company. On October 20, 1937, Dr. Massengill sent a brief telegram to the AMA: "Please wire collect by Western Union suggestion for antidote and treatment following use of Elixir Sulfanilamide." The AMA replied equally tersely: "Antidote for Elixir Sulfanilamide-Massengill not known. Treatment presumably symptomatic." In other words, there was no way to counteract the kidney-destroying effects of the elixir.

The FDA did the best it could with its limited resources to respond to the crisis. It dispatched inspectors to S. E. Massengill's headquarters in Bristol, Tennessee. When they arrived, they learned that the company had already sent a telegram to salesmen, pharmacists, and physicians requesting the return of their remaining supply of the Elixir. This telegram was not exactly an urgent klaxon of alarm, however: "Have withdrawn product elixir sulfanilamide. Please return unused stocks immediately." The FDA inspectors insisted that Massengill issue a more forceful telegram, and on October 19 he sent a new message: "Imperative you take up immediately all elixir sulfanilamide dispensed. Product may be dangerous to life. Return all stocks, our expense."

In the country's first-ever response to a drug crisis, almost every single one of the 239 existing FDA field inspectors was dispatched around the country to recover the toxic medicine. This was a particularly admirable effort, considering that drug safety was not actually part of the FDA's mandate. The inspectors painstakingly tracked down every physician who prescribed Elixir Sulfanilamide, every drugstore that sold it, and every patient who swallowed it. They managed to recover 234 gallons and one pint of the 240 total gallons that had been distributed. The missing six gallons, however, caused more than one hundred deaths.

The media erupted with perhaps the greatest public outrage over business practices since Upton Sinclair exposed the meatpacking industry in his muckraking novel *The Jungle*. When asked about his personal culpability in the scandal, Samuel Evans Massengill declared, "My chemists and I deeply regret the fatal results, but there was no error in the manufacture of the product. We have been supplying a legitimate professional demand and not once could have foreseen the unlooked-for results. I do not feel that there was any responsibility on our part."

Legally speaking, Massengill was correct—based on the existing laws, his company had done nothing that could be considered a felony. The federal court in Greeneville, Tennessee, found S. E. Massengill Company guilty of violating a minor regulation in the 1906 Pure Food and Drug Act that prohibited the labeling of a preparation as an "elixir" if it contained no alcohol. For this violation, the company paid a fine of $150 for each of more than 170 counts of misbranding, totaling $26,000. The families of the 121 victims, however, received nothing.

Harold Watkins, the chemist who formulated Massengill's lethal preparation of sulfanilamide, was not nearly as sanguine as his boss. He was tormented by his role in the catastrophe. While awaiting the federal trial, Watkins shot himself in the head. Samuel

Evans Massengill, on the other hand, stayed on as chief executive; he was the sole owner of the business and could not be ousted. The S. E. Massengill Company continued to operate as a private family-owned pharmaceutical firm until 1971, when it was acquired by Beecham, plc. Beecham merged with another drug company in 1989 to become SmithKline, which merged again in 2000 to become GlaxoSmithKline. Thus, the corporate descendant of S. E. Massengill still endures today, selling billions of dollars of drugs each year.

The massive outcry over the Elixir Sulfanilamide poisonings, including well-publicized letters to President Roosevelt from relatives of the victims, prompted Congress to pass the Food, Drug and Cosmetic Act in 1938 to regulate the sale and marketing of pharmaceuticals. This act established the modern FDA. These days, the Food and Drug Administration oversees the development of drugs from the very beginning, well in advance of any human testing. All pharmacological research that may eventually lead to the development of a commercial drug must be conducted under a system of management controls called "good laboratory practice" or GLP. An executive at the Big Pharma company Cyanamid once told me that GLP was a system "designed to force you to prove that you are not a crook."

Before approving human trials, the FDA reviews a comprehensive dossier of safety testing results from studies performed by the manufacturer in test tubes and on lab animals. If satisfied with this safety data, the FDA approves the manufacturer to proceed with testing on humans, under FDA oversight. A drug is approved for sale only if the FDA determines that it is safe and produces the claimed effects. FDA oversight continues even after the drug is marketed to the public; the agency looks for unexpected or rare reactions to the drug that might have escaped detection during testing.

In 1937, during the height of the sulfa craze, the original FDA had a total field force of 239 inspectors and chemists. In 2013, the FDA had over nine thousand employees and an annual budget in excess of $1.25 billion. As a patient and a consumer, I firmly believe that any industry as potentially hazardous to the public as the pharmaceutical industry needs attentive regulatory oversight. The real question is, what is the proper balance between government regulation and freedom to innovate?

In 1937, there was the wrong balance. Drug companies were given far too much license to take risks at the public's expense. Today things are more complicated. Think about the early days of the AIDS crisis when activist groups like ACT UP petitioned the FDA to loosen its criteria for the clinical testing of potential AIDS drugs. Victim advocates argued that AIDS patients were already dying, so why not give them a remote chance at living by allowing companies to test experimental anti-HIV medications? That was a situation where it seemed to make sense to tilt the balance away from safety and back toward innovation.

Personally, after spending almost four decades in the pharma industry, I believe the vast majority of drug researchers are honest men and women dedicated to finding medicines that will truly help sick people. Despite the public's general antipathy towards Big Pharma, most drug recalls are not the result of deception or greed but authentic mistakes made by people working at the frontiers of what is known about human biology. At the same time, considering the staggering sums of money involved in contemporary drug development, the temptation for cutting a few corners remains high.

I was working for the pharmaceutical division of American Home Products (AHP) during the heyday of the fen-phen diet drug fad. Fenfluramine was first clinically introduced by AHP as a weight-loss drug in the 1970s, but it never became very popular because the weight loss it produced was temporary. The drug sold modestly until 1992, when researchers at the University of Rochester published a study showing that when fenfluramine was combined with a second weight-loss drug called phentermine (also manufactured by AHP), it was more effective at reducing the weight of the chronically obese than dieting or exercising.

The fen-phen cocktail became an overnight sensation. By 1996, 6.6 million prescriptions for fen-phen were being written annually in the USA. Unfortunately, though AHP manufactured both drugs, they had never conducted any experiments on the fenfluramine-phentermine combination. I, along with several of my research colleagues at AHP, argued that the company should launch an effort to develop knowledge about this suddenly popular drug combination. We warned the executives that AHP was now selling something to millions of people that it did not fully understand.

Management brushed aside our concerns. After all, the FDA had approved both drugs and getting FDA approval was neither easy nor economical. Furthermore, the University of Rochester scientists had independently recommended the drug combo as a safe and effective method of weight loss. AHP had already done everything they were supposed to do, the executives argued, so there was no need to spend additional resources on new research or testing. They would soon come to regret their budget-protecting decision.

In 1996, a paper appeared in the *New England Journal of Medicine* reporting on twenty-four patients who had used fen-phen. The authors described a correlation between the use of the drug combo and mitral valve dysfunction. Later that year, a thirty-year-old woman developed heart problems after taking fen-phen for a month. She died. Soon, the FDA received over a hundred reports of mitral valve-related heart disease in patients taking fen-phen. Further investigation revealed that, even though it was very rare for either drug alone to exhibit toxicity, when they were used together the combination was more likely to cause heart problems. Eventually, the FDA determined that fenfluramine was the bad actor in the cocktail, and in 1997 the agency ordered the withdrawal of fenfluramine from the market.

Patients began to sue AHP in droves. *American Lawyer* magazine ran a cover story on fen-phen, observing that more than fifty thousand product liability lawsuits had been filed by alleged victims of the weight loss cocktail. As of 2005, AHP (later renamed Wyeth and now part of Pfizer) was offering settlements of $5,000 to $200,000 to many victims who had sued. These offers were often rejected as too low. Estimates of AHP's total liability have run as high as $14 billion.

The fen-phen debacle points to the difficulty of getting the regulatory balance right. Unlike the development of Elixir Sulfanilamide, there was tight, cautious oversight at every stage of the development of each of the two weight-loss drugs. While AHP had never explicitly tested the fen-phen combination, it was neither unusual nor illegal for physicians to prescribe legitimate drugs in new combinations. Even though AHP management had decided to skip further experimentation on fenfluramine once the drug suddenly became popular, it is not obvious that this decision was ethically objectionable. After all, they had always hoped that the drug would become popular, and the rigorous system of FDA testing presumes that a drug's use might be widespread.

With Elixir Sulfanilamide, both the owner and the chemist at S. E. Massengill were clearly culpable for bypassing even the most rudimentary forms of safety testing. In contrast, while AHP as an organization remains morally and legally accountable to the victims of fen-phen, it is difficult to point the finger at any one person—one greedy villain or one clueless executive—whose poor judgment renders him or her ethically responsible for the drug cocktail–induced injuries. The mitral valve problems were a very rare response that simply did not show up until extremely large numbers of people began extensively using the two drugs in combination.

To my mind, where AHP appears to have crossed the line was in their marketing. While it was entirely legitimate for its salespeople to remind doctors about the University of Rochester study, it was both unethical and illegal for the salespeople to explicitly recommend that doctors prescribe the two drugs together unless the FDA had actually approved the use of the fen-phen drug combination. Despite this, AHP sales representatives frankly encouraged physicians to prescribe the cocktail.

The stories of Elixir Sulfanilamide and fen-phen underscore one of the most troublesome aspects of developing any new drug: side effects. In the case of Massengill's drug, the primary side effect (the lethal destruction of an imbiber's kidneys) resulted not from the drug's active agent but from the formulation—the manner in which the medicine was prepared for human consumption. Today, the regulations of the FDA are designed to ensure the impossibility of a drug company releasing a formulation containing any toxic adulterant.

In contrast, the dangerous side effects arising from the fen-phen cocktail resulted from two different sources: first, an unexpected interaction between the active agents in the two drugs; second, the occurrence of rare side effects from fenfluramine that never appeared in the drug's clinical trials. Side effects from drug interactions remain a fairly common risk today. For instance, combining alcohol with benzodiazepines (such as Librium) or combining MAO inhibitor antidepressants (such as Nardil) with SSRI antidepressants (such as Prozac) can potentially be lethal. Though the FDA monitors a drug's performance after its launch in order to rapidly identify unexpected side effects that may result from drug combinations, it is certainly possible that dangerous or deadly side effects may emerge from future combinations of FDA-approved drugs.

Why do drugs seem to produce so many unwanted side effects in the first place, even when you are only taking one particular drug for one particular reason? To my mind, there are two basic mechanistic explanations. First, many drugs affect multiple physiological targets in the body because different parts of the body often share similar biological targets. A good example are the classic chemotherapeutic agents that attack cancer. "Chemotherapy" destroys cancer cells by acting on the process of rapid cell division in the cancer cells. However, many other cells in the body also undergo rapid cell division (such as the bone marrow cells that create new blood), and they too are negatively impacted by chemotherapy. Another example is Viagra, which targets the PDE5 enzyme in the penis. PDE5 is also present in the cardiovascular system, which is why Viagra causes unintentional flushing and headaches. In addition, an extremely similar enzyme known as PDE6 is found in the retina of the eye, so high doses of Viagra can produce blindness.

Because any given type of receptor in our body usually exists in multiple locations and is often similar to other types of receptors, it is very difficult to find a chemical that affects only one specific physiological target. Sometimes, however, the fact that a drug acts on multiple targets simultaneously can be beneficial. For example, antipsychotic drugs activate multiple targets—but the effects on two of these targets (dopamine receptors and serotonin receptors) fortuitously cancel each other out. The action of an antipsychotic drug on dopamine receptors can generate uncontrollable movements, but the actions of the same drug on serotonin receptors attenuates these movements.

There is another basic mechanical reason that drugs produce unwanted side effects. Drugs are chemicals. Whenever foreign chemicals are introduced into the body, they can interact with our body's natural free-floating chemicals (known as metabolites, the by-products of healthy physiological processes) in undesirable ways. A drug can serve as an imperfect substitute for the metabolites, for instance, causing our body's processes to operate in a flawed manner. A drug can even undergo a direct chemical reaction with our body's metabolites, producing new and possibly toxic compounds.

Very often it is simply not possible for a chemical to produce beneficial effects without also producing unpleasant, harmful, or dangerous effects, so the drug hunter (and the FDA) must always weigh the balance between these positive and negative responses before deciding if a particular drug is suitable for human use.

To find new cures and remedies, we must be willing to take some risk. Without risk, it is simply not possible to develop novel drugs. We can reduce this risk by establishing more regulations, but these regulations impose ever increasing costs on drug development, to the point where today the average cost of developing a new drug has been estimated to be in the range of 1.4 to 1.6 *billion* dollars. This exorbitant financial bar ensures that very few promising drugs will ever make it out of the planning stages. If we want to eliminate the prospect of another fen-phen disaster, the only solution is to expand the regulations for approving drugs to ensure that a wide variety of drug combinations is also evaluated, thereby increasing the cost of developing new drugs even further—which will reduce the number of new drugs even further. This remains the most daunting barrier to drug hunting in the modern era. It is almost inconceivably expensive to safely search for new drugs, but without those exorbitant safety expenses, vulnerable people might get injured or die.

In addition, the FDA remains a government bureaucracy, with certain ineliminable inefficiencies that can disrupt or deter the development of useful drugs. As one example, in the late 1980s one of my colleagues at the pharma company I was working for resigned in a pique of anger at our bosses and took a job at the FDA. This type of career move from Big Pharma to the FDA happens all the time, so I did not think too much about it. I continued my daily work on drug development. But then I started to notice that some of our submissions to the FDA were receiving extraordinary scrutiny. Each time we submitted a report, the FDA found trivial and obviously unintentional errors, and demanded that we revise and re-submit our work. Delays stretched on and on. Even worse, all these re-submissions meant our costs were steadily mounting.

Finally, Cyanamid decided to find out why the FDA was wrapping us up in so much red tape. It was my former colleague. He was using his new job at the FDA to hinder our drug hunting efforts. This was not technically illegal—he was not fabricating his objections and roadblocks out of the air; he was merely seeking out any flaw in our submissions, no matter how irrelevant or miniscule, and then declaring that this flaw required a comprehensive (and costly) correction. No doubt it was spiteful and vindictive. Even so, all we could do was hope that our tormentor would find some reason to get angry at his bosses at the FDA and, with luck, resign from there, too.

Today, the FDA remains our country's greatest protection against another Elixir Sulfanilamide disaster. But this protection comes with a very significant cost. About two weeks after

September 11, 2001, I had to fly from New Jersey to Boston, a common route for me. When I arrived at Newark Airport it was quiet, deserted, and downright eerie. My flight—usually oversold with more than a hundred passengers jostling to get on—had only two dozen passengers. I sank into an aisle seat, and a woman sat down across the aisle from me. A minute later a swarthy man with a full dark beard started trudging down the plane toward us. The woman grabbed my hand and in a shaky, frightened voice whispered, "Oh my god . . ."

Nothing happened, of course. Though the man appeared Middle Eastern, he could have been from many other places and was probably just as anxious about the flight as the rest of us. In such an environment of fear and paranoia, everyone was grateful for the establishment of the TSA, and in the early years after 9/11 we welcomed the reassuring presence of TSA officers in the airports.

But these days, of course, everyone has complaints about the TSA. We have to empty our pockets, remove our shoes, pull off our belts, and haul out our laptops every time we travel. We can no longer carry on beverages or even bring common toiletries like shampoo, toothpaste, or shaving cream, except in miniaturized versions that we always seem to forget. Security queues get ever longer and slower, and we occasionally miss flights because of the prolonged time it takes to reach our gate.

Just as protecting society from terrorism requires a constant rebalancing of safety with individual freedoms and costs (for example in the form of higher taxes or airline fees to pay for the expanded security), protecting society from dangerous drugs requires a constant rebalancing of safety with costs and the delay of vital drugs from reaching the clinic.

7 The Official Manual of Drug Hunting

Pharmacology Becomes a Science

A snake oil salesman

"When I was at Yale, the Harvard and Yale students used to argue as to which school had the worst course in pharmacology."

—Dr. Louis S. Goodman, author of *The Pharmacological Basis of Therapeutics*

During the second half of the nineteenth century, thousands of Chinese workers surged into the United States to build the transcontinental railroad. These immigrants brought with them one of their favorite folk remedies, a greasy extract from the mildly venomous Chinese mud snake. The Chinese laborers rubbed this liniment on their joints to relieve pain from arthritis and bursitis. Many entrepreneurs observed the popularity of the exotic salve among the Asian emigrés and got to wondering whether they might produce their own American version of snake oil.

One of these gumptious capitalists was a man who came to be known as the Rattlesnake King. Clark Stanley, a cowboy, claimed that Hopi medicine men had revealed to him the wondrous power of prairie rattlesnake oil. He peddled his own snake oil concoction at the 1893 World's Exposition in Chicago. His method of promotion demonstrated his understanding of the value of showmanship when hawking a new pharmaceutical product. In front of a rapt audience of potential customers, Stanley reached his hand into a wriggling sack and plucked out a long rattlesnake showing its fangs. He deftly slit it open with a knife and eviscerated it before plunging the serpent into boiling water. As the fat rose to the top of the cauldron, the Rattlesnake King skimmed it off and scooped it into a clear four-inch-tall bottle. Clark Stanley's Snake Oil was snapped up by his enthralled spectators.

In truth, Stanley's Snake Oil usually contained no snake oil whatsoever, rattlesnake or otherwise. Instead, the bottles contained a mixture of mineral oil, beef fat, red pepper, and a dollop of turpentine to give the brew a medicinal smell. Even though Stanley's customers were buying a completely bogus product, it hardly mattered: whether authentic or ersatz, snake oil in all its varieties is therapeutically worthless.

Nearly a half century after Stanley's Snake Oil was marketed to a gullible public at the World's Exposition, the 1937 Elixir Sulfanilamide disaster highlighted the dangers of unregulated medicine, signaling the end of more than fifty years of a Wild West, anything-goes approach to selling drugs in the United States. Yet, even though the elixir deaths marked a momentous shift in society's attitude toward the government's involvement in the pharmaceutical industry, embodied in a much more powerful and active FDA, it did not alter one of the most troubling facts about drug hunting: There was still no coherent science of pharmacology.

As the 1940s dawned, even though consumers were demanding that the government do a better job of monitoring the development of new medicines, there was very little hard science that could be relied upon to guide the FDA's oversight. Not only did the vast majority of medical schools in the 1940s lack a pharmacology department, most did not even offer a pharmacology course. One reason was that there were simply no fundamental philosophical tenets or organizing causal principles in drug science, in the same way that aeronautical science, for example, was organized around the four force vectors of flight, which enabled a practitioner to accurately predict the amount of lift that would be produced by any given wing design. Instead, pharmacology was a chaotic grab bag of ideas from microbiology, physiology, chemistry, and biochemistry, as well as an incoherent casserole of clinical observations about the effects of drugs in various circumstances.

Because of rampant confusion between fact and falsehood in the field of drug development, most physicians thought it pointless to try to teach medical students any principles of pharmacology—after all, with so much confusion, there was as much chance of teaching something wrong as there was of teaching something helpful. Instead, students learned about the nature of drugs directly from the doctors on their own training wards; these were older physicians who simply shared their own experiences with various medications. Thus, guidelines about which drugs to use in which circumstances were highly personal lore handed down from mentor to apprentice, just as it was during the era of medieval apothecaries. It was simply not possible to learn about drugs from books or the scientific literature.

The story of how drug hunting, drug testing, and drug administration finally became a legitimate if unique science originates with two young men at Yale. In the late 1930s, Alfred Gilman and Louis Goodman were newly appointed assistant professors in the Yale Medical School Department of Pharmacology, one of the few such departments in the country, where they were assigned the unenviable task of teaching pharmacology to the medical students. One of the biggest problems that this pair of instructors had to confront was the absence of any useful pharmacology textbook. All the existing textbooks were poorly written or hopelessly outdated; most suffered from both of these shortcomings.

So Gilman and Goodman decided to team up and write their own book. Just as Cordus did five centuries earlier when he penned his groundbreaking opus *Dispensatorium*, the two young scientists set out to create nothing less than a comprehensive compendium of everything that was known about drugs. And, like Cordus, they took a pragmatic and evidence-focused approach to their project, relying on data from published studies rather than oral lore. But they went further than Cordus ever could have, by drawing upon other medical sciences in a highly original attempt at placing what little was known about drugs within the larger framework of what was known about human physiology, pathology, and the principles of treatment. One of their boldest decisions was to structure their book around pharmacodynamics, a nascent field that studied the relationships between the dose of a drug and its physiological effects. Today, pharmacodynamics is a central concept of modern pharmacology, but in the 1930s many of Goodman and Gilman's colleagues believed the field offered little of value. However, Goodman and Gilman wanted to gather together in a single place everything factual and proven that was known about medicines.

Not surprisingly, the textbook proved to be a mammoth undertaking. It quickly began to swallow up all of the young collaborators' time, making it difficult for them to focus on teaching and impeding their ability to do research. This made the book a highly risky venture. Goodman and Gilman's academic careers—including their prospects of getting tenure—were based on publishing original research, not writing a new student textbook. But they pressed on, accumulating an ever more elaborate pharmacopeia—and sending the book's word count higher and higher.

The Drug Hunters, the book before you, contains about 75,000 words. The King James Bible, containing the holy documents of two religions, is 783,137 words. But when the publisher, Macmillan, finally received Goodman and Gilman's completed manuscript, the editor was shocked to see it contained more than a million words.

Macmillan immediately lobbied to reduce the length of the manuscript. The authors refused to cut a single sentence. They believed they had compiled the first comprehensive scientific survey of the science of drugs. Quite reluctantly, Macmillan eventually agreed to publish an uncut edition of *The Pharmacological Basis of Therapeutics* in 1941, but priced the 1,200-page book at $12.50 (about $185.00 in today's money), which was more than 50 percent higher than most medical textbooks of the era. The dubious publisher expected few sales at this outrageous price point, and printed only 3,000 copies. They promised the authors a bonus case of Scotch if the first printing sold out in four years.

Goodman and Gilman got their Scotch, and it took only six weeks. The first edition of the textbook went on to sell more than 86,000 copies. *The Pharmacological Basis of Therapeutics* was instantly embraced by the pharma community as its unifying bible. It contained detailed, evidence-grounded information about every known drug, and more than that, it was the first time that this information was organized around guiding scientific principles that attempted to draw a sense of deeper order from the cacophony of knowledge. For the first time, if you wanted to confidently learn about a particular drug—or if you wanted to teach yourself the entire science of drugs—all you needed to do was delve into Goodman and Gilman. In fact, if the book had any meaningful shortcoming, it was its extreme scholarliness, which often made it a difficult read for the medical students for whom it was originally intended.

As they were publishing their book, Goodman and Gilman went to work for the military as part of America's war effort during World War II, where they implemented a rational approach to drug hunting by incorporating the ideas they had laid out in *The Pharmacological Basis of Therapeutics*. The United States Army contracted Yale University to develop antidotes for Germany's toxic gas weapons, including organophosphate and nitrogen mustard. Gilman and Goodman were put in charge of this countermeasures project, and during their research they observed that nitrogen mustard was cytotoxic, meaning the gas destroyed human cells, especially the fast-growing cells in bone marrow, the digestive tract, and lymphatic tissue. The two young scientists wondered if nitrogen mustard might be repurposed as a lymphoma cancer treatment, targeting the fast-growing lymphoid tumor cells without killing healthy cells.

At the time, the only treatments for cancer of any kind were surgery and radiation therapy. Goodman and Gilman tested nitrogen mustard on a lymphoma-afflicted mouse. Its tumors diminished rapidly. Next, they tested the mustard on a patient in the terminal stage of lymphosarcoma, when radiation therapy no longer worked. The response was dramatic: within two days the patient's tumors had softened; within four days, the tumors were no longer palpable; a few days after that, the tumors had vanished. Goodman and Gilman had invented the very first form of chemotherapy for cancer, an impressive product of rational drug hunting.

Louis Goodman was also interested in drugs that affected the nervous system. One of these was curare, an extract of the bark of a flowering plant that twined around tropical trees. European explorers of the upper Amazon river basin reported that the indigenous peoples hunted their prey using arrows or blowgun darts dipped in curare. (The word *curare* comes from the Carib word *uireary*, which means "to kill birds.") The drug leads to the paralysis of the respiratory muscles and, eventually, asphyxiation. Interestingly, curare is harmless if swallowed, because the compound cannot pass through the lining of the digestive tract into the blood; as a result, South American tribespeople were able to safely eat their curare-poisoned game. Until the 1940s, curare was largely an exotic curiosity in medical circles, but Goodman wondered whether curare could be used as a surgical anesthetic.

Any surgical anesthetic must possess two properties: (1) it must produce unconsciousness and (2) it must block pain. To determine whether curare fulfilled these requirements, Goodman persuaded the chairman of the Utah Medical School Department of Anesthesiology to allow Goodman to inject him with curare and watch what happened. After giving his elder colleague a heavy dose of the drug, Goodman's team proceeded to jab his skin with pins. They also monitored the anesthesiologist's consciousness via a prearranged method of communicating through eye blinks.

Unfortunately, the anesthesiologist blinked his eyes in response to questions, demonstrating that he was fully conscious and violating requirement number one. Even worse, he still felt pain. He mentally recoiled from every prick of the needle, violating requirement number two. In fact, the curare did not alter his consciousness at all; it merely prevented his muscles from moving. Indeed, the dose was too high, and 30 minutes after receiving the injection he stopped breathing. Goodman's drug hunting experiment would have resulted in the death of the chairman of anesthesiology, but fortunately Goodman was able to ventilate the anesthesiologist's lungs using a rubber bag until the drug wore off. This time, Goodman's attempt at establishing a new therapeutic use for an interesting compound ended in failure, but once again the experience proved to him that it was possible to evaluate potential new drugs in a systematic, logical manner.

Today, Goodman and Gilman's tome is longer than ever, but—now in its twelfth edition—*The Pharmacological Basis of*

Therapeutics remains the preeminent pharmacology textbook for twenty-first-century medical students and the bible for all drug hunters. It may also be the only textbook to inspire a child's name. Alfred Gilman named his son "Alfred Goodman Gilman," after the two authors of the history-making book. The pharmacology-tinged name does not appear to have been a handicap; young Goodman Gilman went on to become a professor at the University of Texas Southwestern in Dallas and in 1994 won the Nobel Prize in Medicine for his own drug-hunting research on G-protein coupled receptors, a major class of drug targets.

With the 1941 publication of *The Pharmacological Basis of Therapeutics,* drug hunters finally possessed a coherent framework for a science of drugs. Now all they had to do was use it to find new medicines.

8 Beyond Salvarsan

The Library of Dirty Medicine

The library of dirt

"The earth will open and bring forth salvation."

—Isaiah 45:8

As the world's first bona fide cure for an infectious disease, Paul Ehrlich's syphilis-slaying Salvarsan was hailed as a "miracle drug." There was just one problem. Syphilis was the only disease it cured.

At first, Ehrlich had hoped that his magic bullet might kill other infectious bacteria, too, but experiments in the 1910s showed that the drug had no effect against any pathogen other than the syphilis *Treponeme* bacterium. Other bacteria-based diseases like tuberculosis, tetanus, anthrax, whooping cough, gonorrhea, diphtheria, typhoid fever, strep throat, rheumatic fever, and staph infections all remained untreatable and potentially fatal. Salvarsan was available during World War I, but it was useless to prevent deaths from bacterial infection, which comprised about one third of all the soldiers' deaths.

In 1928, a microbiologist at St. Mary's Hospital in London was studying *Staphylococcus aureus,* a type of bacterium that usually lives quietly and harmlessly on our skin. But if it somehow manages to leak into our bloodstream, watch out. The resulting infection can be as mild as impetigo, a skin disease that produces little blisters or sores on children, but other staph infections can be life-threatening, such as septicemia (blood poisoning) or toxic shock syndrome, a disease so lethal that it can turn a healthy person into a corpse in a matter of hours. This microbiologist was studying staphylococcus cells using the agar plating technique, which means that the bacteria are grown on a dish of nutrients (the agar). The solid surface of the agar plate allows researchers to examine visible colonies of bacteria spread out on the dish instead of peering into the murky stew of a bacteria-saturated test tube.

One day, the microbiologist came into the lab and found something strange. The scientist's name was Alexander Fleming, and you probably know the story of what happened next. According to legend, Fleming left the window to the laboratory open, and when he examined his agar plate he discovered there was fungus growing in it, presumably fungus that had floated in through the window. (I have always doubted this account. I frequently work in laboratories with the windows closed—or without any windows at all—and I often end up with contaminated plates. Fungus spores are always lurking in the air.) Though we don't know exactly where the fungus really came from, Fleming was sure about one thing—the staphylococcus colonies were not growing anywhere near the invading legion of fungus. Fleming guessed that the fungus was producing a substance that was toxic to the staph bacteria. He began to wonder: could this mysterious substance be the basis for another miracle drug?

Fleming named the still-unidentified substance "penicillin" after the fungus that was growing in the dish, *Penicillium chrysogenum*. Next, he conducted a series of experiments to test out the bacteria-fighting properties of penicillin. To his delight, penicillin snuffed out a range of different pathogenic bacteria. Fleming published his optimistic findings in 1929 in the *British Journal of Experimental Pathology*.

Yet, even though Fleming's penicillin exhibited potency against nasty ailments like diphtheria, rheumatoid fever, and strep throat, there were still a couple of obstacles preventing it from being converted into a commercial drug. First, it was not clear how to manufacture penicillin on a large scale. Salvarsan was a synthetic molecule, the product of chemically twiddling a molecule of dye, which meant that it was straightforward to manufacture as much Salvarsan as you wanted as long as you had the necessary base chemicals. But penicillin was produced by a tiny fungus. The only way to get more penicillin was to grow a lot more fungus and then extract the bacteria-fighting compound from the *P. chrysogenum* culture. When Fleming made his discovery, there were no known ways of growing the massive amount of fungus that would be necessary to produce enough penicillin to benefit a small town, let alone all of England. In fact, you could count the number of infected people who would be cured by the existing fungus-growing techniques on a single hand.

Second, Fleming found that penicillin took a long time to stamp out bacteria. We now know that this mistaken conclusion was due to Fleming's faulty method of administration. Instead of dosing his subjects with penicillin through an injection or a pill—methods that deliver the medicine into the patient's bloodstream—Fleming administered penicillin as a topical agent. He chose to rub penicillin on his sick subjects' skin because he was worried that the human body would break down the drug before it could start to work. Fleming's penicillin was further weakened as a consequence of him using low doses, a necessity due to the difficulty of producing the antibiotic.

Because of the difficulty of growing *P. chrysogenum* and the seemingly sluggish effects of the drug, Fleming could not persuade a chemist to help him create a more purified version. Discouraged, Fleming continued to work intermittently on his fungal antibiotic throughout the 1930s, but the medical community ignored his work. They presumed penicillin would never be useful as a commercial medicine. From 1929 to 1940, penicillin remained on the shelf, little more than a laboratory oddity—unused and virtually unexamined. It might never have become one of the most famous drugs in history if a pair of immigrants had not decided to take a second look.

Howard Florey and Ernst Boris Chain were both scientists and both born outside of Britain, but other than that their backgrounds could not have been more different. Chain was born in 1906 to a Jewish family in Berlin. Howard Florey was born in 1898 in Adelaide, South Australia. Chain's father was a chemist who owned a number of chemical factories; Chain followed in his father's footsteps and in 1930 received a chemistry degree from Friedrich Wilhelm University. The Nazis came to power soon after, driving Chain to cross the English Channel to the United Kingdom in 1933 with just £10 in his pocket. Florey, in contrast, studied medicine at the University of Adelaide, where he received a Rhodes scholarship that funded his graduate studies in pathology in England.

In 1939, the Rhodes Scholar and the Jewish refugee joined forces in Florey's pathology laboratory at Oxford to pursue a single mission: testing whether penicillin might actually be useful as a general-purpose antibiotic. After reading Fleming's papers, they speculated that a more purified and concentrated version of penicillin might be more effective at killing bacteria than the diluted and adulterated version that Fleming used. Since Chain was highly trained in chemistry, he set to work preparing a more highly purified version of the compound. When he finished, the two scientists tested their preparation on mice. Their stronger version of penicillin—which today is known as benzylpenicillin—demonstrated that they could cure bacterial infections much faster and more completely than Fleming's formulation ever did. They published their impressive new results in 1940.

On seeing this report, an excited Fleming immediately telephoned Florey to say he would be visiting their pathology lab within a few days. More than a decade had passed since Fleming had published his initial paper about penicillin, so when Chain learned of Fleming's impending visit he remarked, "Good God! I thought he was dead."

In 1941, Florey and Chain treated their first patient. Albert Alexander had scratched his face on a rose thorn. Unfortunately for the ill-fated Alexander, the thorn was infested with malignant bacteria. The scratch became infected and the infection spread rapidly. In a few days, his entire face, scalp, and eyes were severely swollen. His eye soon became so badly infected, that the doctors feared the infection might spread into his brain and kill him, so they performed an enucleation—they surgically removed his eyeball. Even this extreme operation failed to halt the voracious bacteria. Facing death, with no other known treatment, Alexander was the perfect candidate for a penicillin trial.

Florey and Chain administered the drug through an injection directly into Alexander's bloodstream. In less than 24 hours, he started to recover. Unfortunately, Florey and Chain had used their entire supply of purified penicillin on their initial dose, which today we recognize was both too little and too brief to wipe out such an advanced infection. Despite the promising start, Alexander relapsed. Though some of the bacteria had been stopped by the influx of penicillin, the remaining bacteria inexorably continued their invasion. A few days later, Alexander died. Florey and Chain realized that if they wanted to fully test the antibiotic properties of penicillin, they needed to come up with a way to generate more of the compound.

The only known way to make penicillin from the fungus was through "surface fermentation," which meant growing *P. chrysogenum* on agar plates. Florey and Chain filled entire bedpans with agar to maximize the surface area, but even this expanded growth medium would never serve as a scalable way to produce the drug. They decided to conduct all their future experiments on children, since their small bodies required less of the drug. Soon, Florey and Chain were able to show that penicillin was very effective at curing a variety of bacterial infections, as long as it was injected directly into the bloodstream (their benzylpenicillin preparation did not work orally) and as long as the dose was high enough. The need for high doses exacerbated the penicillin shortage even further.

Now that penicillin was proven to be an even better miracle drug than Salvarsan, every hospital clamored for some of the woefully insufficient supply. During the early years of World War II, the best source of penicillin was the urine of patients already treated with the drug, since the active compound is excreted into the urine largely unchanged. As a result, hospital staffs conducted great efforts to collect every drop of urine from their patients in order to recycle their precious contents.

The manufacture of penicillin quickly turned into an exasperating problem of industrial production. England was at war with Nazi Germany and fighting for its very survival. It did not have the ability to divert its limited industrial resources from the desperate war effort to the manufacture of a drug, no matter how important. The Rockefeller Foundation, which had been funding Florey's research, urged him to visit the United States and seek help from England's ally. In July of 1941, Florey flew to New York, where he met with government agencies and private firms. Fortunately for Florey and Great Britain, the United States Department of Agriculture decided to get involved.

The USDA had already been working on fermentation methods to increase the growth of fungal cultures in its Peoria, Illinois laboratory; now the Peoria team set to work looking for ways to increase the growth of *P. chrysogenum*. The USDA scientists eventually made two major contributions. First, they found a strain of *P. chrysogenum* growing on a moldy cantaloupe in a Peoria fruit market that produced far more penicillin than any prior strain of the fungus. Second, they discovered that they could produce much more penicillin much more quickly if they cultured the fungus in deep vats containing corn steep liquor (a cheap byproduct of corn milling) and then pumped air through the fungus-infused liquor (a process known as sparging). Best of all, this deep vat fermentation method was scalable. It finally lead to the industrial manufacture of the world's first expanded-spectrum antibiotic.

A consortium of major American drug manufacturers began to work together and share information on the production of penicillin. These companies—Merck, Squibb, Pfizer, Abbott, Eli Lilly, Parke Davis, and Upjohn—came to be known in the pharma industry as "the penicillin club" and represented the Big Pharma firms of the era. It is an interesting comment on the evolution of Big Pharma that only two of these former giants, Abbott and Eli Lilly, still exist as independent companies. Squibb was eventually taken over by Bristol Myers. Merck was forced to merge with Schering. Parke Davis was once the world's largest pharmaceutical company, but it was absorbed by Pfizer, which is now the largest pharmaceutical company of all time.

During the first five months of 1943, enough penicillin was produced in America to treat about four patients. Over the next seven months, enough penicillin was produced to treat 20 patients. Production methods continued to improve so that by the time the Allies invaded France on D-Day, there was enough penicillin to meet all the needs of the Allied forces. For the first time, wounded soldiers could recover rapidly from infections arising from battlefield wounds. Although Chain later learned that his mother and sister had been killed in German concentration camps, he could know that his research played a meaningful role in defeating the Nazis.

By the end of 1944, the deep vat method of production was finally perfected and Pfizer became the world's largest penicillin producer by churning out enough doses for 100 patients every month. Though penicillin was a true miracle drug, some bacteria-borne diseases remained impervious to it. Perhaps the most dreadful of these diseases was tuberculosis, known as the "White Death" because of the anemic pallor of its victims. In the nineteenth century it was also regarded as "the romantic disease" because the thin, wan, melancholy appearance it bestowed upon the afflicted was often considered a "terrible beauty." Dramatists and poets were drawn to the disease because of its tragic and doleful qualities and because it killed slowly, giving its victims time to tidy up their affairs in life and mend broken relationships before their dramatic demise. The heroines in Puccini's *La Bohème* and Verdi's *La Traviata* die of tuberculosis in each opera's final scene; in *La Traviata,* the curtain falls as the doctor makes his pronouncement of death. Who knows—without tuberculosis, the world's great opera houses might be dark today.

In reality, there is very little about the disease that might be considered romantic or beautiful. The tuberculosis bacteria infect the lungs, where they slowly but ineluctably eat away the air passages, causing its victims to cough up blood as they painfully waste away, growing ever paler and thinner. They appear as if they are being consumed, giving rise to the most common epithet for the sickness: consumption. It is also highly contagious, since the pathogen is readily transmitted to others whenever an infected person coughs, sneezes, or spits. (Anti-spitting laws were originally enacted to fight the spread of tuberculosis and remain on the books in most American municipalities.) When penicillin was invented, the only known treatment was to isolate infected patients in a sanatorium and hope for the disease to spontaneously remit. It rarely did.

The tuberculosis bacterium is a very slow-killing pathogen, which tells us that it is also a very highly evolved pathogen. Newly evolved germs like HIV, SARS, and the Nipah virus tend to kill their victims rapidly. This is a faulty strategy from the pathogen's point of view, the equivalent of ripping up its own meal ticket. Fast-acting pathogens kill their host before they have a chance to spread to many other hosts. In contrast, highly evolved diseases milk their host for as long as possible, giving the pathogen a more prolonged opportunity to infect others. Tuberculosis is one of the most advanced of human diseases and seems to be as old as humanity itself. Even today, roughly one out of every three people on Earth is infected, with a new infection occurring once per second. Fortunately, most cases of consumption do not produce any symptoms, but even so, in 2016 there are fourteen million chronic cases worldwide producing about two million deaths each year.

In 1905 Robert Koch was awarded the Nobel Prize for discovering that *Mycobacterium tuberculosis* caused consumption. Scientists tried Salvarsan and, later, penicillin on the bacteria, but neither antibiotic could touch this unusually hearty and resilient germ. Many researchers suggested that certain breeds of bacteria, like *M. tuberculosis,* simply could not be killed by drugs. But one man held a contrarian view.

Selman Abraham Waksman was born in Priluka, near Kiev, Russia, but immigrated to the United States to attend Rutgers College in New Jersey. He received a bachelor's degree in agriculture in 1915. The growth of crops depends on the interaction of the crop and its soil, including the microbes that inhabit the soil. Waksman became interested in this interaction, particularly in dirt, the rich, dark earth that nurtured crops. He began his research career studying soil—in particular, the bacteria in the soil. Soil microorganisms are essential to breaking down organic matter that falls onto the ground and converting them into the nutrients that plants need to grow. Working at an agricultural school, Waksman hoped that achieving a better understanding of soil microbiology would eventually provide a path to improving crop yields.

In science, the greatest discoveries are often obtained by scientists who started out studying one thing and unexpectedly stumble upon something else. For example, biologist Barbara McClintock set out to understand why kernels of corn were different colors, and ended up discovering one of the most important findings in modern biology—transposons, genetic elements that move from one DNA site to another. Similarly, neurologist Stanley Prusiner was doing his residency when a patient with Creutzfeldt-Jakob (CJD) disease came to see him. CJD is a neurodegenerative disease that is always fatal. At the time, nobody had any idea what caused this strange and incurable disease, since its pathogen had never been identified, but in an exhaustive attempt at helping his patients, Prusiner ended up discovering prions, an entirely new protein-based pathogen previously unknown to science. Both McClintock and Prusiner received Nobel Prizes for their unintended discoveries, and Waksman would eventually receive a Nobel Prize for his.

When Waksman learned about the success of penicillin, a compound produced by a familiar dirt-dwelling fungus, he immediately wondered if there might be other soil microorganisms with antibiotic properties. One group of microorganisms that Waksman had been studying for years is known as the Streptomycetes. These bacteria are so abundant that they produce the distinctive "earthy" odor that we associate with freshly turned soil. In 1939, he decided to investigate whether any of the Streptomycetes might kill bacteria. And not just any bacteria—from the start, Waksman hunted for a cure for tuberculosis, the most destructive of the diseases that penicillin failed to tame.

Waksman already knew how to grow and isolate soil microorganisms, since that was his field of expertise. What he did not know was how to develop an effective assay to test whether any Streptomycete-produced compound could kill the tuberculosis pathogen. Though in principle Waksman could simply grow *M. tuberculosis* in a petri dish, then add the test compound—this was how Fleming had discovered the effects of penicillin—Waksman rightly feared that working with large-scale cultures of living tuberculosis bacteria would be dangerous and could lead to the infection of the entire laboratory staff.

This was ultimately a screening problem. Waksman solved it by screening the Streptomycete compounds on a bacterium known as *M. smegmatis,* a species that is closely related to *M. tuberculosis* but is not harmful to humans. As a bonus, *M. smegmatis* grows much faster than *M. tuberculosis,* making it easier to carry out experiments. Waksman hoped that anything that killed the substitute bacteria would also kill tuberculosis. Fortunately for us all, his hypothesis turned out to be correct.

Waksman's laboratory discovered its first antibiotic candidate in 1940, a compound known as actinomycin. It was very effective against a broad variety of pathogens, including tuberculosis, but Waksman's excitement was short-lived. When they tested actinomycin on animals, it turned out to be far too toxic to be useful as a drug. He returned to screening other Streptomycete compounds. In 1942, his laboratory found another antibiotic candidate that we now call streptothricin. This compound was also a very effective bacteria-killer, and this time when it was tested out on animals, the animals didn't die. Not at first, anyway.

Eventually, Waksman's team learned that streptothricin slowly damaged the kidneys. Animals could tolerate a dose of the antibiotic if it was brief, but if they had to be steadily dosed over an extended period of time, the animals' kidneys failed, killing them. Antibiotics kill bacteria by attacking them when the bacteria are growing; when bacteria are dormant, such as in a spore or cyst state, they cannot be killed by antibiotics. The faster a bacterium grows, the easier it is for an antibiotic to kill it, generally speaking. Unfortunately, the highly-evolved tuberculosis bacterium is extremely slow-growing, which meant that any antibiotic would require a particularly long period of treatment to rub out all the bacteria. Streptothricin would not work, either.

Despite Waksman's double disappointment, the indefatigable drug hunter remained confident that his team would prevail. They continued screening Streptomycete-produced compounds, and in 1943 tested a recently acquired strain of *Steptomyces griseus* that was found in a chicken's windpipe. The team found that this unusual strain produced an antibiotic substance that obliterated a wide range of bacteria, including tuberculosis. They tested it on animals, and to their delight, discovered it was not toxic. They dubbed it streptomycin. Streptomycin was developed by Merck into a commercial product and by 1949 had begun to be administered around the world as the first cure for consumption. In short order, it had saved millions of lives.

In the United States, tuberculosis was particularly rampant among poor immigrants, more than half of whom died within five years of diagnosis. In the late nineteenth century, the best available treatment was considered to be sunshine and fresh mountain air. Sunny sanatoriums sprang up through the country, particularly in the Rocky Mountain states. One of the most popular tuberculosis sanatoriums was the Trudeau Institute, founded in the town of Saranac Lake in upstate New York by one Dr. Edward Livingston Trudeau. Ironically, the Trudeau Institute was in a location that was neither particularly sunny nor mountainous, though it didn't really matter—the therapeutic effects of *any* sanatorium on tuberculosis are fairly negligible.

The introduction of anti-tuberculin drugs provoked a sea-change. Rather than waiting around in a sanatorium hoping that your disease might spontaneously remit, tuberculosis patients could now return home with the promise of an honest-to-goodness cure. Today, tuberculosis patients are treated with a cocktail of anti-tuberculin drugs, similar to the cocktails used to treat HIV/AIDS patients. Currently, the recommended cocktail consists of four antibiotics—isoniazid, rifampicin, pyrazinamide, and ethambutol—that almost always cure the disease when properly administered.

Waksman's Nobel Prize–winning discovery threw open the doors to the library of dirt and set off a mad rush by the pharmaceutical industry. Hundreds of drug hunters began digging through the earth all around the world in the hope of finding new bacteria-murdering microorganisms, launching what is now regarded as the "golden era of antibiotic research." Many antibiotics in our twenty-first-century medicine cabinets were discovered during this golden age, including bacitracin (1945), chloramphenicol (1947), polymyxin (1947), chlortetracycline (1950), erythromycin (1952), vancomycin (1954), and many others.

Florey and Chain's redevelopment of penicillin demonstrated to physicians, scientists, and the general public that antibiotic drugs existed that could completely extirpate pathogens within the body, eliminating all symptoms and ensuring the disease would not be spread to anyone else. It was the Holy Grail of early twentieth century drug hunting: *a cure for infectious disease*. It inaugurated an Age of Dirt, as every major pharmaceutical company had a team devoted to searching through the soil. But penicillin also revealed something else—something extremely vexing. After getting exposed to an antibiotic drug, pathogenic bacteria could change their nature so that the drug would no longer harm them. It was as if the germs could pull on a new suit of armor specially designed to turn aside the pharmaceutical weapons slashing at them.

The first reports of a pathogen developing resistance to penicillin appeared in 1947, just four years after the drug began to be mass-produced. But penicillin wasn't the only miracle drug that stopped being so miraculous. Resistance to tetracycline, another early antibiotic, emerged within ten years of its introduction. Erythromycin resistance took fifteen years to emerge, while gentamicin resistance took twelve years and vancomycin sixteen years. At first, scientists were baffled. Every one of their new wonder drugs eventually lost their potency, like an aging stallion. But soon they realized that the pathogens were *evolving*.

This started one of the greatest battles in pharmacology—really one of the greatest battles in all of medicine—an endless arms race between diseases and cures. The pattern remained very consistent. Drug hunters would unearth some new antibiotic drug.

It would kill pathogenic bacteria for a while. But eventually the fast-reproducing bacteria's genome would mutate just enough so that the drug was no longer effective against it.

Often, pharma scientists could tweak an antibiotic into a slightly different compound (known as an analog) that would also kill the mutated pathogen, but eventually the pathogen would mutate again and the analog would no longer work, either. Antibiotic resistance remains an unsolved problem to this day. We still confront many antibiotic-resistant strains of bacteria that have become or are becoming just as lethal as they were before the discovery of penicillin, including *Staphylococcus aureus* (MRSA), *Neisseria gonorrhoeae* (gonorrhea), *Pseudomonas aeruginosa* (pneumonia and sepsis), *Escherichia coli* (E. coli), and *Streptococcus pyogenes*. There is even a new strain of *Mycobacterium tuberculosis* that can survive the standard anti-tuberculosis cocktails.

With the threat of lethal bacterial infections remaining quite real, you might be surprised to learn that in the 1980s Big Pharma started to give up on developing new antibiotics. Why would they abandon a product with such an obvious market? Because antibiotics do not offer a particularly profitable business model for drug companies. Big Pharma prefers drugs that need to be taken over and over again, such as pills for high blood pressure or elevated cholesterol. Medications for such chronic conditions must be taken every day of a patient's life, which can drive tremendous sales. But antibiotics are only taken for a week or so, after which the patient is cured and does not need the drug again. This sharply limits profits.

But the economics of antibiotics are even worse than their one-off method of treatment. As physicians began to realize that every new antibiotic would eventually cause pathogens to develop resistance, they started to stash away each new antibiotic drug to hold in reserve. They only brought these drugs out to use on patients with terrible infections caused by antibiotic-resistant bacteria. This was a sensible way to preserve the potency of new antibiotics, but it meant even more diminished sales for each new (highly expensive) antibiotic a pharma company managed to develop, since doctors would hoard the drug instead of prescribing it.

In 1950, virtually every pharma company had an antibiotic research unit. By 1990, many of the largest American pharmaceutical companies had marginalized antibiotic research, or dropped it entirely. That same year witnessed a sudden resurgence of interest in antibiotics within the scientific community, triggered by the outbreak of MRSA and other antibiotic-resistant germs. The pharmaceutical industry did not embrace this renewed interest, however, continuing their steady withdrawal from the fight against infectious disease. In 1999, Roche pulled out of antibiotic discovery. By 2002, Bristol-Myers Squibb Company, Abbott Laboratories, Eli Lilly and Company, Aventis, and Wyeth had all terminated or severely downsized their antibiotic programs. Pfizer, one of the last holdouts, shut down its main antibiotic research center in 2011, perhaps signaling the twilight of the Age of Dirt. Today, fifteen of the eighteen largest pharmaceutical companies have abandoned the antibiotic market entirely.

I am one of the youngest people alive to have conducted research in a classic Big Pharma antibiotic discovery program—the source of my adventures wheeling around the Chesapeake in my lime-green Monobacvan. It was the dying days of searching for new soil in the hope of unearthing some previously undiscovered bacteria-butchering microorganism. I never did find a new antibiotic in the Delmarva dirt, and even if I had, it quite likely would have been shelved before it got developed into a commercial drug.

Today, things have reached a dangerous state of affairs. Dr. Janet Woodcock, director of the Center for Drug Evaluation and

Research at the FDA, recently stated that, "We are facing a huge crisis worldwide not having an antibiotics pipeline. It is bad now, and the infectious disease docs are frantic. But what is worse is the thought of where we will be five to ten years from now." More than 23,000 people die in the United States each year from a bacterial infection that was once easily treated with antibiotics but that has now developed resistance. That's more than the number of Americans who die from (virus-borne) AIDS each year.

Alexander Fleming made one of the greatest discoveries in human history: a single cure for many diseases. Sadly, this cure is not imperishable. It must be constantly refreshed and renewed, the remedy as dynamic and ever-changing as the blight itself.

9 The Pig Elixir

The Library of Genetic Medicine

Frederick Banting, Charles Best, and Dog 408

"Peace comes from within. Do not seek it without."

—Buddha

For most of our species' existence on our vast green planet, drug hunters searched through the variegated library of plants for new unguents and balms. Botanical medicines were found in abundance. In comparison, the stingy library of animals was a paltry source of drugs; one simple reason is that there is far less fauna than flora on our Earth. Nevertheless, from ancient times until the modern era, humans have extracted myriad drugs from animals. A handful actually worked. Most, however, provided no benefit at all—other than the placebo effect.

Take the rhinoceros horn. It is a common misconception that powdered rhinoceros horn was used as an aphrodisiac or a cure for cancer in traditional Chinese medicine. In truth, no Chinese medical text mentions such uses. Instead, traditional Chinese medicine promoted the rhinoceros horn as a treatment for fevers and convulsions, though it possesses the same power to cure these conditions as it does to cure cancer: none. In fact, a recent monograph entitled *Chinese Herbal Medicine: Materia Medica* compares the consumption of powdered rhinoceros horn to the consumption of fingernail clippings.

Even so, the misbegotten notion that the Chinese used the rhinoceros horn as an aphrodisiac has driven sales of the rare rhinoceros horn in Vietnam and other Southeast Asian countries. This demand has spurred rhinoceros poaching to the point where the International Union for Conservation of Nature now lists three of the five known rhinoceros species as critically endangered.

A similar situation exists for tiger parts. The bones, eyes, whiskers, and teeth of *Panthera tigris* have been used in traditional Chinese medicine for the treatment of a wide range of ailments, including malaria, meningitis, and bad skin. Traditional Chinese medicine claims that almost every part of the tiger can be used medically. Claws—a sedative for insomnia. Teeth—a treatment for fever. Fat—a cure for leprosy and rheumatism. Nose leather—a balm for superficial wounds and insect bites. Tiger eyeballs—a remedy for epilepsy and malaria. Whiskers—an analgesic for toothaches. Brain—an antidote for laziness. Penis—can be ground and stewed into a love potion. Tiger dung—a panacea for hemorrhoids. As you can probably guess, there is absolutely no evidence that any of these preparations have any medical value.

And just as happened with the hapless rhinoceros, the misguided faith in the therapeutic power of tiger pills, powders, and wines has resulted in a catastrophe for the graceful feline. Of the original nine subspecies of tigers, three have become extinct in the last eighty years. Four of the remaining subspecies are considered endangered, two critically so. The International Union for Conservation of Nature estimates that the total population of the six remaining subspecies is under four thousand individuals. (In comparison, there are more than forty million domestic cats in the United States alone.)

But even though the library of plants produced a few ancient

Vindications that survived into the twenty-first century—including morphine, ergot (a drug that is still in clinical use but that has been largely been replaced by new, superior medicines such as the triptan drugs to treat migraine headaches and oxytocin in labor and delivery), and digitalis (still used to treat heart conditions)—not a single pre-twentieth-century drug from the library of animals has made it into modern medicine. Why are there so many more useful medicinal compounds in plants than in animals? We do not know for certain, but one theory is that plants have been defending themselves against insects for hundreds of millions of years, and so plant immune systems produce a dazzling variety of compounds designed to repel, injure, or kill an extremely wide range of predatory bugs. These defensive compounds (which botanists call phytotoxins) are highly bioactive, since they are designed to influence or impair the physiology of insects. Even though human physiology is far more sophisticated than the physiology of beetles and moths, our bodies still share some of the same fundamental biochemistry. Thus, even if a particular phytotoxin does not have the exact same effect on our body that it does on an insect, the compound may still provoke some kind of effect within our own physiological processes—an effect that on occasion may be beneficial to us. Perhaps animals tend to produce far fewer substances with the potential to disrupt physiological processes because they have less of a need to fend off insects or other nibbling creatures, though a small number of animals do produce toxins designed to disrupt the physiology of predators or prey, such as poisonous snakes, scorpions, and toads. Similarly, soil microorganisms have been at war with one another for eons, and so they produce an impressive array of antifungal and antibiotic toxins that can be harvested for drugs.

By 1900, the consensus of the biomedical community was that medicinal preparations from animals were simply not useful, and pharma companies and drug hunters alike had abandoned all attempts at searching animal parts for beneficial compounds. Yet, twenty years after the turn of the century, one of the most important drugs in history was discovered within the organs of dogs.

The story behind this animal-borne Vindication commences in 1897, when Bayer first sold Aspirin to a grateful public and shoveled up undreamed-of profit. The global success of this synthetic remedy opened up a whole new world of pharmaceutical opportunity as drug companies realized that enormous revenues were waiting to be had for any truly original drug. As the twentieth century dawned, many large drug companies began to set up their own drug discovery units to search through the library of molecules for new therapeutic compounds, and one of the first American pharmaceutical companies to try to develop their own drugs was Eli Lilly.

The company was founded by Colonel Eli Lilly, a Civil War veteran and pharmacist, in 1876 in Indianapolis. Most of Lilly's early products were hand-rolled pills coated with sugar, elixirs, and syrups, including the top-selling Succus Alterans, a useless formulation sold as a treatment for syphilis and "certain types of rheumatism and especially skin diseases like eczema, psoriasis, etc." Josiah Lilly took over the family business in 1898 after the death of his father, and eventually Josiah's grandson Eli (named after his great-grandfather) became company president and chairman of the board. Eli Lilly, a third-generation pharma executive, looked with envy upon Bayer's success developing new drugs in Germany and decided that his own company should get into the drug hunting game.

In 1919, Lilly recruited a scientist named Alec Clowes to serve as a kind of roaming opportunist whose job was to sniff out new product opportunities, similar to the contemporary role of a licensing director. Clowes's background was in cancer research. He had spent eighteen years at the prestigious Roswell Park Memorial Institute in Buffalo, where he made a name for himself as an exceptional scientist. He also possessed an entrepreneurial streak that made him attractive to Lilly as the right person to move the company out of drug repackaging and into drug discovery. In 1919, Clowes began reviewing various diseases and ailments to see which ones might be offer the best opportunities for drug development. He quickly settled on a malady without any known treatment: diabetes.

During the second millennium BC, Indian physicians observed that ants were attracted to the urine of certain patients; Egyptian manuscripts from roughly the same era, meanwhile, describe some patients as suffering from "too great emptying of the urine." These are the oldest known records detailing the symptoms of diabetes. The Indians called it *madhumeha,* or "honey urine." The Greeks called it diabetes, meaning "passing through," referring to the excessive discharge of urine. In 1675, a British physician termed it diabetes mellitus, adding the Latin word for "sweet tasting." Today, this form of diabetes is most commonly known as type 1.

Type 1 diabetes almost always starts in childhood and—without treatment—is inevitably fatal. Its victims are usually unquenchably thirsty and insatiably hungry. Nevertheless, even though they consume prodigious amounts of water and food, they slowly lose weight and steadily waste away. Diabetes also impairs blood circulation and damages nerves. Poor circulation often causes blindness when there is not enough blood getting to the retina, and can even cause the loss of limbs. Simultaneously, as their nerves are slowly destroyed, victims feel escalating levels of pain.

When Clowes joined Lilly, most individuals with diabetes perished within a year of being diagnosed. Four thousand years after the "honey urine" disease was first documented, there was still no known remedy. Thousands of plant-based compounds were tried on diabetics during the Age of Plants, all without effect. The Age of Chemistry had also failed to produce any viable treatment. But Clowes hoped to change that.

Fortunately, there was one widely accepted notion about the kind of drug that might be able to treat diabetes, an insight that emerged purely by happenstance. In 1889, two European physicians, Joseph von Mering and Oskar Minkowski, were conducting a series of experiments to determine the function of the mysterious oblong organ located between the stomach and small intestine, an organ known as the pancreas. Their methodology was simple. They removed the pancreas from a healthy dog and watched what happened. And what happened was that the housebroken dog began to urinate on the laboratory floor. All day long.

The researchers knew that frequent urination was a symptom of diabetes, so they tested the dog's urine. It was high in sugar. Von Mering and Minkowski speculated that they had just created the first artificially induced example of diabetes by removing the dog's pancreas. Next, they tried to determine what a pancreas was actually doing that apparently prevented diabetes in healthy individuals. They proposed that the dog pancreases produced a hormone that controls how the body metabolizes glucose, a hormone we now call insulin.

Glucose is used by cells as a prime source of energy. Insulin acts as a kind of key that unlocks a special door on the cell membrane, allowing glucose to enter hungry cells. In the absence of insulin, glucose builds up to high levels in the blood, but the sugar molecules cannot enter cells to feed them. After a while, the high level of glucose overwhelms the kidneys' ability to reabsorb it and the excess sugar starts to spill out into the urine, producing the characteristic "honey urine."

Based on Von Mering and Minkowski's pioneering work, scientists speculated that it should be possible to treat type 1 diabetes patients by giving them insulin. Initially, drug hunters presumed that all they needed to do was remove a healthy pancreas, grind it up, extract the insulin, and inject it into a diabetic person. But harvesting useful insulin turned out to be a near-impossible task. The reason it was more difficult than anybody expected was because of an odd physiological complication unique to the pancreas. One of the two main functions of the pancreas is producing hormones, including insulin. But the other function is to produce enzymes that the small intestine uses to digest proteins. Unfortunately, insulin is a protein. Whenever researchers ground up a pancreas in the hope of extracting insulin, they inevitably mixed together the insulin protein with the protein-digesting enzymes, destroying the insulin.

Despite this daunting obstacle, the medical consensus about insulin remained firm. If you could somehow figure out a reliable way of getting insulin, you would have a cure for diabetes. Scientists all around the world began to investigate different ways of extracting insulin from animals, all without success. One man who came late to the quest for insulin was Frederick Banting.

Born on a farm in Ontario, Canada, Banting had a slow start to his medical career. In 1910, he enrolled in the General Arts program at the University of Toronto but failed his first year. Nevertheless, he managed to get admitted into the University of Toronto medical program in 1912. When Canada was drawn into World War I in 1914, he tried to join the army as a medic. He was rejected. He applied again and was rejected again due to poor eyesight. He applied a third time and was finally accepted, perhaps because of the great need for army medics. He started serving the day after he graduated, but after the war ended, he encountered more professional difficulties. Though he landed a residency at the Hospital for Sick Children, he was unable to secure a permanent spot once his residency ended. He had no choice but to set up his own private medical practice, but he was not successful at that, either.

After a career full of disappointments and failures, Banting had become highly sensitive to perceived professional slights, a quality which would haunt him as he switched to a new career track. After reading a 1920 scientific paper describing an experiment in which the pancreatic duct was tied off, Banting became interested in the quest for insulin. The pancreatic duct is a tube that delivers the digestive enzymes to the small intestine, and the article reported that when the duct was clamped shut the enzyme-producing cells in the pancreas died—but there was a kicker. The insulin-producing cells remained alive and functional.

After reading the article, Banting guessed that if the enzyme-producing cells in a duct-clamped pancreas were no longer producing enzymes, it would finally be possible to safely harvest insulin. It was a pretty good idea—so good, in fact, that it had already been tried by other research teams. Those previous attempts, however, had always resulted in failure. Banting was completely unaware of these prior failures. Inspired by the possibility of curing diabetes using what he wrongly believed was his own original approach, he abruptly made the decision to switch from being a full-time physician to a full-time drug hunter.

In order to pursue his dream of extracting insulin, he recognized that he would need a fully-equipped laboratory where he could work. So he visited the lab of a world famous physiologist at the University of Toronto, J. J. R. Macleod. Macleod listened to Banting's proposal with tremendous skepticism—unlike Banting, he was well aware of all the failed attempts at extracting insulin—but he was ultimately swayed by Banting's passion and drive. Since

Macleod was about to leave Toronto for the Scottish Highlands for his summer holidays, he figured it would not hurt to let Banting try out his ideas while he was away. Macleod generously provided Banting with laboratory space and even appointed a medical student to assist him.

During the scorching hot Toronto summer of 1921, Banting and his young assistant, Charles Best, initiated their experiments in clamping off the pancreatic ducts of dogs. This turned out to be excruciatingly difficult surgery, one of the reasons that previous research teams had failed in their efforts to isolate insulin. The first dog Banting and Best operated on was killed by an overdose of the anesthetic. The second dog died from loss of blood. The third perished from an infection. Seven dogs eventually survived the procedure, but the ligation process remained very tricky. If they tied the suture too tightly, it caused infection. If they tied it too loosely, the enzyme-producing cells would not shrink and die. Five of the seven dogs that survived the operation still produced insulin, but the enzyme-producing cells never atrophied. When they re-operated on these five dogs in a second attempt at clamping off their pancreatic ducts, two more died from complications.

Banting and Best were now halfway through their planned research program and had nothing to show for their efforts. Plus, they were running short on dogs. They combed the Toronto streets for stray dogs, which they snatched up and brought back to their lab, where the unlucky canines were rewarded with invasive surgery. Three weeks later, Banting and Best were finally able to harvest the first atrophied pancreas from a successfully ligated dog. They ground up the pancreas and injected the extract into a test dog, which they had made diabetic through the removal of its own pancreas. At last, success! Within an hour, the dog's blood glucose level had declined by almost half.

They painstakingly repeated this experiment on other diabetic dogs. Though not every dog responded to their insulin treatment, enough did respond to convincingly demonstrate that Banting and Best had produced a feasible treatment for diabetes. Yet even though this triumph was both thrilling and gratifying, their process of extracting insulin remained highly unreliable, often failing to produce any insulin at all. Moreover, each surgically sacrificed dog only produced enough insulin for a few doses. The method certainly could not create enough insulin to help a single human diabetic, who would need to take several doses of the drug every day for the rest of their lives—and the idea of treating the country's entire population of diabetics using the makeshift method of extracting insulin from dogs was downright farcical.

In fact, there was no precedent at all for Banting's novel extraction process. Until now, all commercial drugs were either extracted from plants or created through synthetic chemistry. Banting and Best had invented an unprecedented way to extract a useful drug directly from the bodies of animals. Yet if they wanted to harness this process to deliver enough medicine to treat even a handful of diabetes patients, they had to somehow ramp production up to an industrial scale, while their process barely worked on a micro scale. (You might also consider the disconcerting fact that the only way to get enough insulin to save a diabetic child's life was by slaughtering many, many mammals.)

When Macleod returned from Scotland in the fall, he was astonished to discover that the amateur scientist and young medical student had somehow become the world's first researchers to successfully isolate insulin. Macleod grasped the problem of large-scale production and immediately recognized they needed someone who could optimize the insulin-extraction process. He recruited James Collip, a highly regarded biochemist at the University of

Toronto, to join the project. Collip applied state-of-the-art biochemistry techniques to refine the insulin extracted from dogs into a more purified form.

You might think that Banting would have been delighted by this turn of events. They were on the verge of possessing a legitimate treatment for one of humanity's oldest and most deleterious diseases. Instead, after a lifetime of professional failures, Banting viewed Collip as a rival who was butting in to steal credit. In fact, he viewed Collip with such cynical disregard that he frequently stoked fights with him. Sometimes these fights turned physical. On one occasion, Banting became so enraged at Collip's involvement that the contretemps devolved into a fistfight. Collip ended up with a black eye.

By late 1921, the uneasy team of Banting, Best, Collip, and Macleod had discovered a reliable—though still nonscalable—method of extracting insulin from the pancreas of dogs and had shown that this insulin could be used to successfully treat diabetes in dogs. But if they hoped to demonstrate its effectiveness in humans, they would need a way to scale up the extraction process still further—a prospect that looked increasingly dim in the face of Banting's belligerence toward anyone he perceived as trying to swipe his glory. And this is where Eli Lilly comes into the story.

When Alec Clowes was put in charge of looking for new drug development opportunities at Eli Lilly, he knew that insulin stood a good chance of becoming a blockbuster—if someone could figure out how to make it on an industrial scale. In 1921, Clowes attended an academic conference at Yale where he heard Banting's first major public presentation on his work. As Banting shared their promising results, Clowes was filled with a rising sense of excitement. When the presentation ended, Clowes immediately sent a three-word telegram back to Lilly in Indianapolis: "This is it."

Banting had a very different reaction. He did not like the way Macleod had introduced him to the audience before his talk, speaking in a restrained manner that seemed to reserve all the credit for Macleod himself. He did not like the way all the scientists rushed to Macleod after Banting finished speaking, posing their questions to Macleod instead of Banting. He left the meeting feeling disappointed, angry, and aggrieved, convinced that once again other people were swiping credit for his hard work.

Before leaving New Haven, Clowes left a note at Macleod's hotel saying that Lilly wanted to collaborate with his team to develop commercial insulin production. But Macleod, a Canadian, was reluctant to get involved with an American pharmaceutical company. He had been hoping to work with Connaught Laboratories, a vaccine-producing company affiliated with the University of Toronto. He declined Clowes's offer.

George Clowes was not going to take no for an answer. During the next four months, he made four trips to Toronto to press his case with Macleod. At each meeting, Macleod insisted that he wanted to keep the development of insulin within Canada, while Clowes talked up the advantages that Lilly could bring to the project. Macleod might have been able to maintain his resolve if not for the fact that his research team was falling apart.

During the first few months of 1921, the relationships among the team members were deteriorating fast, and their interactions with the Connaught scientists were only adding to the friction. Much of the strife was driven by Banting's jealous fear of losing credit for and control of a project he still considered his own. By early April, things had gotten so bad that Macleod finally succumbed to Clowes's relentless advances. He wrote to Clowes informing him that they were on the verge of perfecting an insulin isolation method that could be used for commercial production at a new site—preferably a site far away from Toronto and the squabbling team.

Macleod began negotiations for licensing insulin production to Lilly. Clowes quickly arranged for Lilly's Indiana site to obtain large quantities of pig and cow pancreases for the anticipated collaboration. Meanwhile, the Toronto team began looking around Toronto General Hospital for a diabetic patient who could serve as their first human guinea pig. They found Leonard Thompson, fourteen years old and a mere sixty-five pounds. The emaciated boy had been suffering from diabetes for three years and was starting to slip into a coma. Death always followed a diabetic coma. Since Thompson would soon die anyway, the team felt that it would be justifiable to test the insulin on him. But the pilot test got stalled by an unexpectedly contentious question: whose hand would actually get to inject Leonard Thompson with the insulin?

The maneuverings were fierce. Banting, of course, thought he should be the one to perform the injection since he viewed himself as the sole inventor of the insulin-extraction method. However, the director of the teaching wards at Toronto General Hospital where Thompson was a patient refused to allow it. The director insisted that a physician with expertise in the treatment of diabetes perform the injection, not Banting. He selected an intern working under his supervision to conduct the historic injection. Banting exploded. He was being blocked from participating in the first test of his great discovery, and instead some random young intern who had nothing at all to do with the discovery was granted the honor.

Banting demanded that he be the one to wield the syringe. As a rather peculiar compromise, the director of the teaching wards agreed to allow the intern to inject an insulin preparation made by Banting and Best, rather than the more highly purified material from Collip's process. This way, even if Banting's fingers were not actually on the syringe, the material in the syringe could rightfully be said to have been the direct result of Banting's personal efforts. Though this managed to satiate Banting's furor, it turned out to be a big mistake.

Banting and Best's concoction produced only a marginal improvement in Thompson's health. Worse, it provoked an allergic reaction, probably due to the presence of contaminants mixed in with the unpurified insulin. Banting's insistence on using his unrefined version of insulin had made the life of a suffering young boy even worse. The team immediately decided to try Collip's highly purified version instead, while there was still time. This time, it worked. Thompson's blood sugar levels declined dramatically and he began to regain his strength and energy. His hunger and thirst diminished, and he began putting on weight again.

It was the first time any human diabetic patient had been treated successfully.

Thompson continued receiving (purified) insulin injections and, even though insulin was not a cure for the disease, managed to live another thirteen years. Previously, a child was lucky to survive a whole year after getting diagnosed with diabetes. Today, with daily doses of insulin, diabetics can expect to live a full life that is only ten years shorter than a non-diabetic's life, on average. Clowes realized that Eli Lilly had just acquired a true blockbuster. It was the dream scenario for any pharma company: there were more than ten thousand diabetics in the USA—and new ones every year, since type 1 diabetes strikes one in four thousand children—and they would each need to take the drug over and over again for their entire lives. All Lilly needed to do was manufacture insulin on a large scale. But how do you ramp up production when the only known way to manufacture a drug is by growing it inside a living pancreas?

Clowes estimated that it would take at least a year to fully develop a commercial insulin process. Lilly budgeted $200,000 to cover the development costs (about $2.5 million today). Collip and Best immediately departed for Indianapolis, where they explained to the Lilly chemists everything they knew about purifying insulin. Within weeks, the Lilly chemists had replicated their small-scale method. The first industrial-scale run was completed just two weeks after that, producing one hundred times more insulin than the Toronto team's method. Soon, the Lilly insulin plant was running three shifts around the clock, involving over more than hundred scientists. Within two months, the insulin yield increased dramatically . . . but its potency declined. Each forward step seemed to be accompanied by a backward step. It took almost two years, but in late 1922, Lilly finally established a reliable process that produced potent insulin on an industrial scale.

In 1923, insulin was offered for sale to diabetic patients throughout North America for the first time. Though the Canadian drug company Connaught owned the rights to sell insulin in Canada, Lilly was granted exclusive rights in the United States. It was not only a pharmaceutical revolution but a revolution in medical practice—a revolution of the syringe. Even though the hypodermic needle had been invented in 1853, it had always been the exclusive domain of trained physicians. But now, insulin treatment required that patients inject themselves, since type 1 diabetics typically require three or four injections per day, too many for a visit to the doctor. Ordinary kids—and ordinary parents of kids—were taught how to administer the protein drug on their own.

Though Eli Lilly's drug worked, the insulin produced by cows and pigs is not identical to human insulin. As a result, bovine and porcine insulin can sometimes trigger allergic reactions in patients. Some individuals develop a rash, though the most common response to animal insulin was lipoatrophy, a loss of subcutaneous fat. The solution, of course, would be to use authentic human insulin. But how do you get it? The only known way to get insulin was to take a pancreas and grind it up—and there were not too many human volunteers willing to offer up their organs. For more than five decades after insulin went on sale, diabetic patients were stuck using animal insulin and frequently having uncomfortable allergic reactions.

But in the 1970s—a half-century after Banting extracted insulin from a dog pancreas—a new opportunity emerged. In 1972, Paul Berg, a Stanford University professor who studied viruses, performed one of the most important experiments of the twentieth century. He removed a piece of DNA from a cell of bacteria and inserted it into the DNA of a monkey cell. He accomplished this by attaching the bacteria DNA to a harmless virus that Berg used as a kind of Trojan Horse to penetrate the monkey cell's defenses and deliver the bacteria's genes directly into the monkey's genome. This process is known as "recombinant DNA" since it combines the DNA of two different organisms—the bacteria and the virus.

Why was this experiment so important? Because once the monkey cell took up the foreign DNA, the bacteria's genes could begin to produce the same proteins that they would inside a bacteria cell. In other words, the bacteria's genes could co-opt the machinery of the monkey cell and re-directed it to manufacture new molecular products. For drug hunters, it was the opposite kind of recombinant DNA that was so promising—taking the genes out of a mammal cell and inserting them into a bacterium. In 1975, a rabbit gene that creates hemoglobin became the very first mammalian gene to be transferred into another organism when it was inserted into an *E. coli* bacterium in a Petri dish. These bacteria cells could now be engineered to produce rabbit hemoglobin, marking a watershed moment in genetics—and the birth of genetic medicine.

That same year witnessed one of the first and most important conferences about the spectacular advances in recombinant DNA, the Cold Spring Harbor Symposium on Quantitative Biology. I remember my thesis advisor returning from the conference and sharing what he had learned with great excitement. "Any human gene can now be harnessed to make human proteins in a test tube," he gushed. "And the place to start is obvious. We should clone the insulin gene and use it to make human insulin."

The insulin gene turned out to be a great choice for the early attempts at developing genetic medicine, and not just because of the huge demand for insulin. The gene for insulin is extremely short, and the smaller a gene the easier it is to manipulate. In 1976, Herb Boyer (a biochemistry professor at the University of California, San Francisco) and Robert A. Swanson (a venture capitalist) started a new company in San Francisco to develop drugs using the new recombinant DNA technology. Genentech's very first project was to manufacture human insulin.

This was an entirely new approach to drug hunting. Instead of searching for new molecules in plants during the Age of Plants, or searching for new synthetic tweaks to existing molecules during the Age of Synthetic Chemistry, or searching through the soil for new bacteria-fighting compounds during the Age of Dirt, Genentech was searching through the human genome for fragments of DNA that could produce useful protein-based drugs. But even though the library of potential genetic drugs was new, the drug hunting story was the same as always. Finding useful drugs is very, very hard, and gets even harder the longer the search goes on.

It took Genentech more than a year to isolate the human insulin gene, and by then they were running out of money fast. They needed a new financial partner to help continue funding their drug development, a partner who could supply enough cash to allow the insulin project to reach a commercial conclusion. There were two obvious targets for Genentech to potentially partner with: Eli Lilly and E. R. Squibb. Even in the late 1970s, Lilly remained the unchallenged leader in insulin production, controlling about 95 percent of the United States insulin market. Squibb, in contrast, was a much smaller niche player who controlled the remaining 5 percent of the market. Genentech's executives reasoned that Squibb might be the better choice, since Squibb would presumably be interested in boosting their undersized market share and recombinant human insulin offered an unprecedented opportunity to do exactly that.

Genentech reached out to Squibb and proposed teaming up. Though Squibb was a large pharma corporation with a huge research staff, they had no experience at all with recombinant DNA technology. So Squibb did the same thing that all Big Pharma companies do when they do not understand some new science: they hired a consultant. Sir Henry Harris, Regius Professor of Medicine at Oxford University, was a pharma consultant with a very impressive resume. Harris had been trained as a physician before spending an admirable career researching tumor cells. Unfortunately, none of Harris's experiences in the biology lab qualified him to evaluate the Genentech proposal, which dealt with cutting-edge genetic technologies that he had never encountered before. But if Harris lacked the apposite expertise, he did not lack in self-confidence.

Harris reviewed the details of Genentech's proposed method for manufacturing human insulin in bacteria cells and rendered the following analysis: Proteins are three-dimensional molecules. The precise three-dimensional shape of any particular protein matters greatly to the operation of physiological processes that use the protein. The specific amino acids that compose any molecule of protein can combine in many different geometries to form many different shapes, but for a given protein to function properly in the body it must be in the proper shape for the body to recognize it and make use of it. So far, so good, but now Harris made an unjustified claim.

He insisted that if human insulin genes were put inside bacteria cells, the bacteria would manufacture insulin protein molecules in a different three-dimensional shape than when the molecules were manufactured in human cells. Since it would be impossible to re-configure the geometry of the improperly-shaped insulin molecules, Genentech would never be able to produce true human insulin. Thus, Harris told Squibb to pass.

Squibb took Harris's considered opinion very seriously and, on the basis of his advice, rejected Genentech's collaboration request. Genentech was surprised by Squibb's reaction, but the Squibb executives refused to believe Genentech's earnest attempts to show why Harris's thinking was wrong. After all, all Genentech could offer was a promise that they would be able to solve the geometry problem, without any concrete evidence that they could.

So Genentech turned to Eli Lilly.

Lilly appraised the situation in an entirely different way. They recognized that there was a small but meaningful chance that Genentech would be able to produce human insulin. If Genentech succeeded and Lilly was not involved, the economic consequences would be catastrophic for Lilly. Insulin was one of the most important market franchises for Lilly—they basically owned the only known treatment for diabetes—and they simply could not risk the possibility of losing the entire market, so matter how unlikely that potential loss. Thus, in 1978, Lilly agreed to collaborate with Genentech.

Sir Henry Harris's analysis of the difficulty of creating properly-shaped proteins using recombinant DNA technologies turned out to be wrong. He guessed that insulin produced by human genes inserted in *E. coli* would have the wrong shape. He was right about that. But in short order Genentech solved this seemingly intractable problem. It developed a biochemical process by which the improperly-shaped insulin harvested from *E. coli* was successfully re-folded into its correct shape. Using Lilly's funding, Genentech produced the first batch of human insulin, manufacturing the precious protein inside a test tube. In 1982, human insulin went on sale for the first time. Today, virtually all insulin is produced using recombinant DNA technologies—and Eli Lilly remains the world leader in insulin production to this very day.

Sir Henry Harris's erroneous opinion had an outsized influence on the course of my own career. In the late 1970s, I was very intrigued by all the new developments in recombinant DNA, and I was eager to use the new technology in my own drug hunting. But I was hired by Squibb in 1981, just after Harris had delivered his pessimistic pronouncement about genetic medicine, and my manager informed me that the company had absolutely no interest in using recombinant technologies to make protein drugs. Squibb missed out on one of the most momentous revolutions in the history of drug hunting—and so did I. Instead, I was assigned to use molecular biology methods to develop traditional drugs, and that is exactly what I did for the rest of my career.

While I was hunting for conventional drugs in the library of dirt and the library of synthetic chemistry, there was a mad rush through the shelves of the newly opened library of genes. Pharmaceutical companies raced to identify new therapeutic proteins they could grow inside bacteria. Since many hormones are proteins, many of the earliest efforts focused on pharmaceutical hormones. After the tremendous success of recombinant insulin, the next recombinant protein to go on sale was human growth hormone (HGH), a treatment for dwarfism, in 1985. Genentech manufactured HGH because the hormone was relatively easy to produce using recombinant technology, even though the market for HGH is far smaller than that for insulin. HGH was followed by interferon, a cancer treatment, by Biogen in 1986; erythropoietin by Amgen in 1989 to treat kidney failure; and clotting factor VIII by Genetics Institute in 1992 to treat hemophilia A.

At first, Big Pharma was thrilled by the realization that recombinant technology granted them virtually unlimited ability to cure any disease caused by a missing protein. But this initial flush of excitement soon dimmed. The fact of the matter is that there are simply not that many diseases caused by missing proteins. By the early 1990s, after producing about a dozen new recombinant drugs, the industry was running out of diseases to treat. Drug hunting followed the same trajectory it always did: a new library of potential molecules is discovered, there are a few major discoveries, the entire industry swarms through the library, and in short order the library is exhausted. Of course, there always seems to be a new library to search and the biotechnology industry soon found another one, known as recombinant monoclonal antibodies.

Here's how monoclonal antibodies work. In response to the presence of a pathogen, the human body's white blood cells produce antibodies—chemicals that attack the invading bacterium, virus, fungus, parasite, or other foreign agent. But since every pathogen is different—sometimes extremely different (just consider the difference between, say, athlete's foot fungus and tapeworms)—each pathogen is vulnerable to a different type of antibody. Thus, if we want to kill an intruder, our body needs to produce the right kind of antibody. Or, even better, produce multiple kinds of pathogen-specific antibodies, which each inflict different damaging effects on the target. Our white blood cells accomplish this using a very sophisticated process. When a germ is detected, the white blood cells (specifically, the B cells) start to rapidly reproduce, but each child cell is a different variant of the parent white blood cell. The body can produce literally millions of white blood cell variants in a very short period of time. Each of these variants produces a different kind of antibody. Thus, we might say the body uses a just-in-time "weapons on demand" system: if it detects an enemy jet fighter, it produces different types of anti-aircraft missiles; if it detects an enemy tank, it produces different kinds of anti-tank rockets; if it detects enemy soldiers, it produces different types of guns.

If a drug hunter thinks that a particular type of antibody might make a useful drug, then he can put human white blood cells into a petri dish and manipulate them into producing the specialized white blood cells that manufacture the desired antibody (typically this is done by exposing the white blood cells to a substance that will trigger the formation of the requisite specialized cells). Next, the drug hunter can isolate the specialized cells that produce the target antibody and then, using recombinant DNA methods, extract the specific genes from these cells that are responsible for creating the antibody, and then use these genes to churn out as much of the antibody as he needs. Finally, he can take this antibody and convert it into a useful drug. Antibodies produced in this fashion are known as monoclonal antibodies because the compounds are from one highly specific type of white blood cell (*mono-clonal* means "single-branch"). The library of monoclonal antibodies is now a mainstay of recombinant DNA drug development and has generated drugs for diseases ranging from multiple sclerosis to rheumatoid arthritis.

The Nobel Prize committee recognized the historic research effort that led to the original development of animal-based insulin by awarding the 1923 Nobel Prize in Medicine to Frederick Banting and J. J. R. Macleod. As you may have guessed, Banting did not react to this award with delight or pride. Instead, he was furious that the Nobel committee forced him to share the prize with Macleod. In Banting's mind, since he had come up with the basic idea for extracting insulin from dogs by clamping shut their pancreatic ducts, he deserved all the credit, even though without Macleod providing him with a laboratory, an assistant, a biochemist, and credibility, Banting would never have been able to bring his vague (and unoriginal) idea to fruition. Banting refused to attend the Nobel award ceremony Stockholm and just stayed home.

I wish I could tell you that drug hunters are a courtly and charitable bunch, but Banting illustrates one of the most enduring facts about my profession. Every successful drug hunter is as distinct as the drug he (or she) discovers.

10 From Blue Death to Beta Blockers

The Library of Epidemiological Medicine

John Snow's map of cholera

"Superior doctors prevent the disease from happening, mediocre doctors treat the disease before fully evident, inferior doctors treat the disease after it is apparent to everyone."

—Huang Dee Nai-Ching, 2600 BC

Cholera is a particularly nasty disease of the small intestine whose prime symptom is "rice water," a watery diarrhea redolent of fish. A victim may excrete up to five gallons of diarrhea each day. Vomiting and muscle cramps are common. The resulting dehydration is often so drastic that a victim's electrolytes become unbalanced, debilitating the heart and brain. Cholera is called "the blue death" because the afflicted's skin can turn bluish-gray from the extreme loss of fluids. Without treatment, about half of the disease's victims die.

Throughout the nineteenth century, wave after wave of cholera pandemics swept through Europe and much of the world. The second wave decimated Ireland in 1849, killing off many of those who had been lucky enough to survive the Irish potato famine. The epidemic then washed up on American shores via ships crammed with Irish immigrants, eventually infecting President James K. Polk. The disease swept west and extinguished some six thousand to twelve thousand travelers along the California, Mormon, and

Oregon trails, mostly pioneers hoping to make their fortune in the California Gold Rush before the ill fortune of rice water put an end to their dreams. As this virulent wave was finally winding down, a brand new globe-hopping wave of cholera came bursting out of India and smashed into London in 1853.

The Blue Death claimed the lives of more than ten thousand Londoners in a single year. One man who became obsessed with the terrifying intestinal disease was an English physician named John Snow. The son of a coal laborer, Snow had grown up in poverty in one of the poorest neighborhoods in York, where his family's ramshackle house was flooded every time the nearby River Ouse overflowed its banks—and it overflowed often. Snow was working as an anesthesiologist at St. George's Hospital in London during the new pandemic, and on August 31, 1854, he took charge of treating cholera patients in the Soho district where he lived. Over the next three days, 127 Soho residents died. By the end of the following week, three quarters of the entire population of Soho had fled, rendering the vacant neighborhood a ghost town. By the end of the next month, out of the scant citizenry left behind, another five hundred had died, with far greater losses throughout the rest of England. Snow later called it "the most terrible outbreak of cholera which ever occurred in this kingdom."

Nobody had the vaguest idea what caused cholera or what its risk factors might be. The London outbreak occurred seven years prior to Louis Pasteur's publication of the germ theory of disease and forty years before Robert Koch (who won the Nobel for showing that a bacterium caused tuberculosis) finally convinced the medical community that cholera and other diseases were truly caused by germs. Snow set out to investigate the cause of one of humankind's most loathsome scourges without any knowledge of infectious pathogens, at a time when the leading explication of disease was miasma theory.

Miasma theory—the notion that disease was caused by "foul air"—certainly seemed like a plausible explanation for cholera. Blue Death was prevalent in many lower-class neighborhoods, locales that inevitably featured the reek of animal and human feces blended with the wet stink of rotting garbage. Another popular opinion held that the supposed moral depravity of the lower classes somehow weakened their constitutions and made them more vulnerable to disease. Snow was skeptical of both miasma theory and depravity theory, however. He suspected, instead, that there might be something in the water. But if you do not know about germs or have the technology to detect them, how can you confirm that water hides some kind of contagion?

Snow approached his investigation in a wholly original manner—so original, in fact, that it would lead to the founding of an entirely new field of medicine. He scrutinized a map of the Soho neighborhood and began to systematically document where each case of cholera had occurred. (Today, this area is the Carnaby Street shopping district in Westminster.) At each location where an afflicted Soho resident lived, he drew a short black bar, stacking multiple bars perpendicular to the adjacent street. He drew 578 bars in total. Next, he marked the location of each public water pump in the neighborhood. London's water supply consisted of a system of shallow public wells where people could pump out water to carry back home. These wells were supplied with water through a jumble of water lines controlled by about a dozen water utilities. As convoluted as London's water system was, the city's sewage system was even more chaotic and ad hoc. Personal privies emptied into cesspools, cellars, or an unfathomable tangle of sewer pipes.

Worst of all, the London aquifer allowed for the easy mixing of the liquids in the cesspools and wells.

Snow noticed several interesting features on his marked-up map. Even though a large Soho workhouse just north of the Broad Street pump housed over five hundred paupers, very few of its residents came down with cholera. Similarly, not a single person who worked at a brewery one block east of the Broad Street pump contracted the disease. Nevertheless, despite these two anomalies, Snow's map made one thing perfectly clear: most of the cholera deaths occurred in residents who lived near the Broad Street pump.

Convinced that whatever was responsible for the disease must be coming out of the well on Broad Street, Snow approached the local city council and demanded that they remove the pump. The council was skeptical. How could that well be polluted? After all, they pointed out, the water from Broad Street was clear and tasted better than most pumps in Soho. In fact, many people avoided the water in their local well and traveled all the way to the Broad Street pump to obtain its clean water, especially residents who lived near the smelly Carnaby Street pump.

But Snow persisted. He showed that the workhouse near Broad Street, where almost no residents got sick, had its own independent well. He pointed out that the workers at the brewery near Broad Street, where nobody got sick, were allowed to drink all the beer they wanted, and he suspected something about the beverage prevented the disease (during brewing, the beer wort is boiled for an hour, killing most bacteria). Perhaps most tellingly, he observed that all the residents near the Carnaby Street pump who got sick were the very ones who traveled to fetch the supposedly clean water at the Broad Street pump.

Eventually, the council relented and granted him permission to shut down the well. Snow immediately removed the handle from the Broad Street pump, making it impossible to get water . . . and that was the end of the cholera epidemic in Soho.

We now know that the Broad Street well was contaminated with the pathogen *Vibrio cholera,* a bacterium that infected residents with every gulp. Yet, even without this knowledge, Snow's original method of investigation—focusing on both geography and population—provided an effective means to control the disease. This was the first scientific example of epidemiology, the study of patterns of diseases in the population. Today, John Snow is regarded as the father of epidemiology.

In some sense, Snow was quite lucky. Unlike an experiment (which can demonstrate cause and effect), an epidemiological study cannot prove causality. It can only demonstrate a relationship—in Snow's case, a relationship between the location of victims and the location of water pumps. It could have been the case that something other than the water or the water pump was causing the disease; there was no way to know for sure solely using Snow's map. Though Snow's conclusion—that there was a contaminant in the Broad Street well—turned out to be correct, epidemiological studies are more susceptible than experiments to misleading findings.

As one example, epidemiological studies in the 1930s found an extremely high correlation between the consumption of refined sugar and the incidence of polio. Does eating sugar cause polio? Not at all. Polio is caused by a virus that is transmitted through drinking water, similar to cholera. So what is the connection to refined sugar?

Babies are born immune to polio because they inherit protective antibodies from their mothers. However, these antibodies wear off after a few months. If you are exposed to polio while you still possess your mother's antibodies you will not get sick. Instead—and rather remarkably—the infection induces your immune system to produce its own antibodies, which will then protect you against polio for the rest of your life.

If, on the other hand, you are exposed to polio after your mother's antibodies wear off, you will develop the full-blown disease. These individuals still produce their own antibodies after the infection strikes, but in many cases this happens too late to protect them from the worst ravages of the disease—lifelong paralysis. Thus, if you contract polio as an infant, you will experience a barely noticeable infection. If you contract polio as a young child or adult, you will suffer devastating effects. In poor countries with inferior sanitation, almost everyone is exposed to polio during their infancy. No problem; they still have mom's antibodies. But before the development of a polio vaccine, in developed countries with excellent sanitation, people usually did not encounter the polio virus until they had reached late childhood or adulthood. Catastrophe.

So what is the sugar connection? In the 1930s, when the epidemiological study was conducted, citizens of wealthy countries (with superior sanitation) enjoyed the luxury of eating refined sugar. In contrast, people in poor countries (with dismal sanitation) could not afford refined sugar. Correlation, not cause and effect.

On the other hand, epidemiology can also generate powerful new insights that overturn conventional medical wisdom—and can lead to unexpected new opportunities for drug hunting. A good example of this is a doctrine-shattering revelation about hypertension divulged by one of medicine's most famous epidemiological studies. You probably know that hypertension—high blood pressure—is unhealthy and merits treatment. But until the 1960s, many physicians actually held the opposite view, a conviction reflected in the antiquated medical term "essential hypertension": for decades, the medical establishment thought that hypertension was *essential* for maintaining good health. John Hay, professor of medicine at Liverpool University, expressed the prevailing sentiment when he wrote in 1931, "there is some truth in the saying that the greatest danger to a man with a high blood pressure lies in its discovery, because then some fool is certain to try and reduce it."

Doctors believed that hypertension was a kind of natural compensatory mechanism that kept the heart pumping properly. President Franklin Delano Roosevelt was a lifelong hypertensive patient, but his doctors feared that it might be dangerous to lower his blood pressure, so they left it alone. FDR died from a stroke during his fourth term—almost certainly the consequence of untreated hypertension. But the fallacious notion of "essential hypertension" was finally disproved by the longest running epidemiological study of all time, the Framingham Heart Study.

Launched in 1948, the Framingham Heart Study initially tracked 5,209 residents of the town of Framingham, Massachusetts, a small working-class municipality about twenty miles west of Boston. Its mission was (and still is) the identification of risk factors associated with the development of cardiovascular disease, one of the leading killers in the 1940s. It was the Framingham Heart Study that first demonstrated the effects of diet and exercise on the prevention of heart disease.

Like John Snow, the original Framingham investigators were skeptical about the prevailing medical theories of their time. Most doctors believed that heart disease was a natural consequence of aging and, as such, finding a medicine to prevent heart disease would be as likely as finding the Fountain of Youth. In contrast, the Framingham scientists speculated that cardiovascular health was influenced by lifestyle and environment. They hoped that a large-scale epidemiological study might identify these lifestyle and environmental factors and lead to new methods of intervention to reduce the risk of cardiovascular disease and strokes.

The investigators knew that the study would need many years before it would be possible to draw firm conclusions, and consequently the first reliable findings were not reported until the early 1960s, more than a decade after the start of the Framingham Heart Study. Among other findings, they showed that stroke was correlated with three separate physical conditions: clogged arteries (atherosclerosis), elevated serum cholesterol (hypercholesterolemia) . . . and hypertension.

Since the Framingham study—like all epidemiological research—was correlational rather than causal, it was not clear if hypertension actually *caused* strokes, or if there was some other shared cause that produced both hypertension and strokes, in the same way that eating refined sugar and contracting polio both resulted from first-world lifestyles in the 1930s. Some doctors critical of the Framingham Heart Study, for example, suggested that both hypertension and strokes were inevitable side effects of aging. But there was an unexpected source of support for the startling idea that "essential hypertension" might not be so essential after all—a drug known as Diuril.

In the 1950s, Merck boasted a program to hunt for compounds that inhibited an enzyme called carbonic anhydrase. These inhibitors reduced blood acidity, a common medical condition that often resulted from kidney or lung problems. For good health, the acidity of our blood must stay within a very narrow range or else we experience headaches, dizziness, or exhaustion, or—if our blood acidity rises particularly high—we might slip into a coma. Carbonic anhydrase inhibitors helped restore blood acidity to a healthy level, but these drugs also provoked an unanticipated side effect. They made patients pee. Physicians call such pee-inducing drugs diuretics.

Increased urination can lower blood volume, and therefore can lower blood pressure. (When there is less fluid circulating in your blood, your heart does not need to work as hard to pump blood through your body, which reduces your blood pressure.) Thus, Merck's carbonic anhydrase inhibitor drugs not only reduced blood acidity (Merck's original objective)—they unintentionally reduced hypertension, too.

Of course, at the time there was no perceived medical need to reduce hypertension. But since Merck now had in their possession a set of diuretic drugs, they looked for other reasons that a patient might want to increase her rate of urination. They soon identified another use for their carbonic anhydrase inhibitor drugs: helping patients who suffered from edema. Edema refers to a swelling of the body produced by the abnormal accumulation of fluid beneath the skin and in the cavities of the body. For example, pulmonary edema is a swelling of the lungs that often occurs when the heart becomes too weak to effectively pump blood out of the lungs, which causes fluid to accumulate in the air spaces in the lungs. Merck realized that carbonic anhydrase inhibitors would be a useful treatment for pulmonary edema, since lowering a patient's blood volume through increased urination would (1) reduce the amount of fluid available to pool around the lungs, and (2) reduce total blood volume, making it easier for the heart to pump blood out of the lungs.

It was at this moment that serendipity struck. While Merck was working to find the most potent and efficacious carbonic anhydrase inhibitor, they stumbled upon a drug that did *not* inhibit carbonic anhydrase yet was an even more powerful diuretic than their existing drugs. They eventually named it Diuril. They had no idea how it worked, but when Merck tested Diuril on patients suffering from pulmonary edema they found it be safe and extremely effective. Thus, the hunt for a drug to treat blood acidity led to an entirely new kind of drug that treated pulmonary edema, a completely different condition. But that was not the end of the story. A Merck scientist named Karl Beyer thought that Diuril could be used for yet another purpose—to "treat" hypertension.

Of course, at the time, the idea of treating hypertension was akin to how we might consider "treating" yawning—why would you want to mess with something so natural and normal? Even so, there was a minority of physicians who suspected that high blood pressure was dangerous rather than a marker of good health. Beyer quietly passed Diuril to his colleague Bill Wilkerson, a physician, and asked Wilkerson to slip the drug to a few hypertensive patients to see what happened. As expected, their blood pressure went down. Beyer knew then that Diuril could be the first clinically effective anti-hypertensive drug—but there was not yet a market for such a medication. When Diuril went on sale to the public in 1958, its primary use was as a treatment for edema.

Nevertheless, other drug companies noticed that Merck had created an effective anti-hypertensive drug and—fearing that they might lose out on some unknown future market opportunity—tried to develop their own. This led to an entire class of Diuril copycat drugs known as the thiazides that served as anti-hypertensives; within a few years after Diuril came out, more than six thiazides had been approved by the FDA.

At first, these diuretic anti-hypertensive drugs were not used very often. But then the first round of the Framingham Heart Study came out, showing a link between hypertension and stroke. Even though many physicians reacted with skepticism to this finding, other doctors knew that safe and effective drugs—the thiazides—were available to lower blood pressure and decided that prescribing these drugs to their patients with high blood pressure presented a favorable risk to reward ratio. If the Framingham link between hypertension and stroke was actually due to cause and effect, well, the thiazides would probably reduce their hypertensive patients' chances of a stroke. If the link was *not* causal, on the other hand, the physicians calculated they would be doing little harm by prescribing the thiazides. The FDA also supported the prescription of the various anti-hypertensive drugs to patients with high blood pressure, since the FDA realized that the only way scientists could establish a causal link between hypertension and stroke (instead of the correlational link in the Framingham Study) was by actually reducing hypertensive individuals' blood pressure and observing what kind of effect it had—by conducting an ad hoc experiment, in other words.

The Centers for Disease Control and Prevention, which monitored the incidence of stroke in the national population, soon noticed that there was a clear reduction in the number of people who were having strokes—and determined this reduction was due to the increase in patients taking anti-hypertension drugs. The medical establishment quickly changed course and began recommending that high blood pressure should be treated. "Essential hypertension" became "unhealthy hypertension." This was one of the first cases where epidemiology and Big Pharma worked hand in hand to overturn conventional wisdom and lead to a dramatic shift in attitude toward a major medical condition, saving countless lives. The incidence of stroke in the US declined by almost 40 percent between the years 1955 and 1980.

Now that it was clear that anti-hypertensive drugs were beneficial, drug hunters embarked on the search for the perfect anti-hypertensive drug. The thiazides were only modestly effective in lowering blood pressure and had one manifestly undesirable side effect—frequent urination. If someone could come up with a way to reduce blood pressure in a more effective manner—without unpleasant side effects—there would be tremendous profit potential, since patients who needed anti-hypertensives would have to take them every day of their lives. And there was such a someone—a man by the name of James Black.

James Black was an unlikely drug hunter. Born in 1924 in the small Scottish town of Uddingston, Black was an excellent student and studied medicine at the University of St. Andrews. Unfortunately for Black, by the time he graduated he had accumulated large debts. He had little choice but to take the best paying job available, which was a teaching job at the University of Malaya in Singapore. He was finally able to return to Scotland by joining the faculty at a veterinary school. He attempted to make the best of his unfavorable professional situation and began to study adrenaline's effect on the human heart, particularly in those suffering from angina.

You are likely familiar with adrenaline's role in the fight-or-flight response—if you encounter something dangerous, like a stranger with a gun, you experience a surge of adrenaline that makes you hyper-alert and ready for action. But adrenaline also serves another physiological role, as a hormone that regulates our blood pressure. Thus, Black concluded that any drug that blocked adrenaline should also lower blood pressure. Armed with this promising idea, Black approached the British company ICI Pharmaceuticals in 1958 and applied for a job as a drug hunting scientist. Despite the fact that Black was a veterinary professor without any pharmaceutical experience, he had an excellent reputation as a researcher and was hired by ICI, where he quickly set to work looking for compounds that might block the effects of adrenaline.

It was known that there were two different types of adrenaline receptors, known as alpha receptors and beta receptors. Studies revealed that the beta receptors were the ones involved in the regulation of blood pressure. Black surmised that if he could block a person's beta receptors, he would be able to reduce her blood pressure. The challenge, though, was to figure out how to block the beta receptors without blocking the alpha receptors, which were molecularly quite similar and which controlled other physiological functions unrelated to blood pressure. Black set to work on finding a compound that would differentiate between the two types of adrenalin receptors and in 1964 discovered propranolol, a drug that selectively blocked beta adrenaline receptors. This was the world's first anti-hypertension "beta blocker."

Propranolol reduced blood pressure without the diuretic effects of the thiazides. It rapidly became one of the best-selling drugs in the late 1960s and 1970s and was prescribed throughout the world. For his groundbreaking work, Black received the Nobel Prize in Medicine in 1988.

Even though the beta blockers were a clear improvement over the thiazides, they still harbored two major flaws. Beta adrenaline receptors are also found in the lungs, where they regulate the size of the airways. Blocking the beta receptors in the lungs causes the airways to constrict. (Indeed, many modern inhalers used to treat asthma attacks contain drugs that *stimulate* the beta receptors in the lungs.) Thus, propranolol and other early beta blockers featured a very untoward side effect—they made it more difficult to breathe. Treating an asthma patient with beta blockers can be downright dangerous. In addition, beta blockers elicit another side effect in men that may be far less physically risky than constricted airways but can be just as psychologically harrowing: impotence.

Thus, every available anti-hypertensive exhibited some meaningful shortcoming. The Holy Grail of anti-hypertensive drugs remained elusive. But I worked at the place where it was finally found. In the early 1980s, during my first pharmaceutical industry job with Squibb, I became acquainted with two snake charmers, Dave Cushman and Miguel Ondetti. They were drug hunters at Squibb who happened to be very interested in the venom produced by pit vipers. The venom of these snakes knocks out their prey by drastically reducing their victims' blood pressure, rendering their victim unconscious. Cushman and Ondetti reasoned that it should be possible to isolate the compound in pit viper venom that reduced blood pressure and convert it into an anti-hypertensive drug.

They started by studying one of the venom's most active components, a substance known as teprotide. They discovered that teprotide inhibited an enzyme in the body known as ACE (angiotensin converting enzyme). Even though adrenaline contributes to the regulation of blood pressure, we now know that ACE serves as the true "master controller" of our blood pressure. In effect, by inhibiting ACE, snake venom shuts down the body's ability to control blood pressure, and without this control, blood pressure dropped.

Cushman and Ondetti set out to develop teprotide into a usable drug. The two men formed an unusual drug discovery team, since they could each seem like yin to the other's yang. Cushman was a loquacious pharmacologist, energetic and mischievous, while Ondetti was a methodical chemist who was serious and reflective. Cushman was passionate about science but perhaps equally passionate about comic books. If you saw him at the copy machine, there was a 50 percent chance he was copying a scientific paper and a 50 percent chance he was copying a new comic book to share with the department. Despite their antithetical personalities, they were extremely effective collaborators.

They started out by seeing if they could somehow use pure

teprotide as a medicine with little chemical modification. They quickly ran into a severe problem: teprotide is not orally active. The digestive enzymes in the stomach destroy the compound, which makes sense, since a pit viper's venom is produced in the snake's mouth and it would be catastrophic if the snake swallowed an orally active poison. If teprotide could only be administered through injections, it would be necessary to inject the drug several times a day, every day. Just thinking about this unending ordeal might release enough stress hormones to counteract any benefits from the drug!

Since neither prey nor patients enjoy injections, Cushman and Ondetti knew they needed an orally active compound if they were ever going to create a commercial version of the drug. They began synthesizing molecules that were similar to teprotide but which they hoped might withstand the rigors of the human stomach. At this stage, it is customary to evaluate thousands or even tens of thousands of molecules—this is the inevitable trial-and-error screening that all drug hunters must go through. But Cushman and Ondetti synthesized and tested only a few hundred molecules by taking a novel approach to the process of screening.

The two scientists understood the biochemistry behind the action of the ACE enzyme and thus could predict the types of compounds that would likely inhibit the enzyme. As a result, the earliest compounds they synthesized worked fairly well, and then they proceeded to further tweak these compounds based on their insights into the likely activity of different molecular structures. After each tweak, they tested the new compound to gauge its effects and evaluate the accuracy of their guesses. If drug screening is usually like randomly spinning the reels on a slot machine, then Cushman and Ondetti's approach was more like figuring out the internal mechanics of the slot machine and then pulling the lever when the machine was about to make a payout.

Using this approach (now called "rational design"), Cushman and Ondetti rapidly synthesized a very effective ACE inhibiting compound, which they dubbed captopril. The rational design process was another landmark in the history of drug hunting. Paul Ehrlich was the first person to come up with a wholly original approach to designing a drug from scratch when he conceived of loading a toxin on a molecule of dye, though he still relied on blind trial-and-error to test a number of possible toxic warheads. In contrast, Cushman and Ondetti came up with another original approach to designing a drug from scratch, but instead of relying on blind trial-and-error, they used their knowledge of chemistry, biochemistry, and human physiology to make a series of increasingly effective educated guesses that allowed them to find what they wanted in record time and with minimal cost.

Captopril blocked the action of ACE, thereby reducing blood pressure. But it was also orally active, able to survive the corrosive acids of digestion. With their potent new drug in hand, you might think that Cushman and Ondetti were now on the glide path to drug hunting triumph. Ah, but the life of a drug hunter is never easy.

The executives at Squibb were hesitant to approve the next stage of drug development, testing the efficacy and safety of captopril, because they would require large-scale—and very expensive—clinical studies. Squibb was already selling a popular beta blocker, nadolol. The business people argued that captopril could cannibalize nadolol sales, so that any potential profits from captopril would come at the cost of reduced nadolol sales. They calculated that the net annual sales gains from captopril would be—at most—in the low hundreds of millions of dollars. Though this might seem impressively high (particularly in the 1970s), it was still not enough to justify authorizing the exorbitant expenses of clinical studies and marketing. Squibb decided to keep captopril on the shelf.

Cushman and Ondetti were very disappointed, to say the least, but since they were scientists they asked management for permission to publish their findings so that they could at least claim credit for their breakthrough research. Most pharma companies are extremely reluctant to let their scientists publish anything that might give competitors an advantage, but Cushman and Ondetti pointed out that captopril was securely patented and therefore publishing presented little risk to the company. Eventually, Squibb management consented to the drug hunters' request—a decision probably motivated in part by feelings of sympathy for having terminated their highly original drug.

Cushman and Ondetti published the details about captopril in several leading pharmaceutical journals. Instantly, the academic medical community sat up and took notice. They recognized that the two Squibb scientists had discovered an entirely novel means for controlling blood pressure. Soon, a number of prominent doctors from major medical schools were reaching out to Squibb. These physicians assumed that Squibb would soon be starting clinical trials on the exciting new drug and they wanted to get involved and run trials at their own schools.

Such overwhelming interest from academia in a shelved drug was unprecedented, and the executives at my company huddled together and finally decided to give captropril the green light. The ensuing clinical trials demonstrated that captropril was a fantastic and safe anti-hypertensive drug. The FDA approved it in 1981. During its first full year of unrestricted commercialization, captopril generated more than a billion dollars in sales. It was so profitable, in fact, that it made more money for Squibb than the rest of its drug portfolio combined.

You might expect that landing a blockbuster drug would have sent Squibb's stock price soaring. But for some reason that I (and my bosses) never quite understood, Squibb's stock price was slow to reflect the soaring sales of captopril. Squibb's competitor Bristol Myers noticed the disconnect, however, and swooped in and purchased Squibb at a discount price. Thus, in a truly ironic twist, Squibb's demise as an independent company was caused by one of the first blockbuster drugs to be developed out of epidemiological research.

11 The Pill

Drug Hunters Striking Gold Outside of Big Pharma

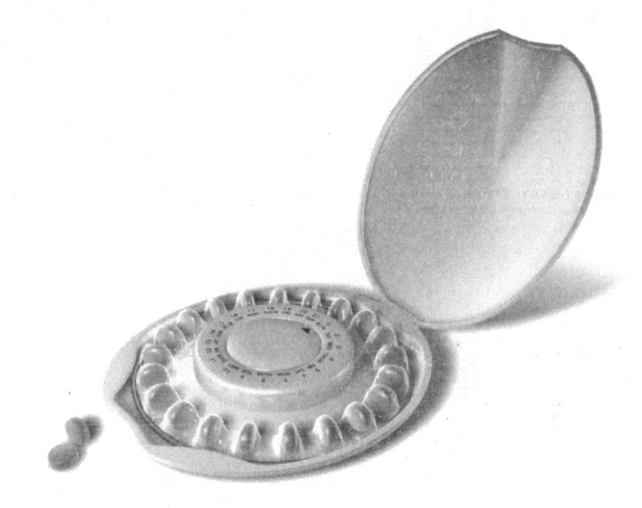

The birth control pill

"No woman can call herself free who does not own and control her body. No woman can call herself free until she can choose consciously whether she will or will not be a mother."

—Margaret Sanger, *Woman and the New Race*, 1922

We have ventured through the Age of Plants, the Age of Synthetic Chemistry, the Age of Dirt, and the Age of Genetic Medicine, each era marked by the opening of a new library of molecules to explore and a new generation of drug hunters rushing in to seek their Vindication. But on rare occasion, a medicine is discovered outside the major libraries, far from the well-funded labs of Big Pharma. Sometimes these stories of "independent" drug hunting cross continents and decades, involve an eclectic and strange cast of characters, wobble through mistakes, misconceptions, and misfortune, and then, at the end of it all, produce a drug that changes the history of the world. And if we wished to single out the independently developed drug that probably did more than any other to change the basic social fabric of modern civilization, it is the drug so prominent and influential that it has come to be known as "the Pill."

The 1970s were the golden age of disaster films, when Hollywood churned out suspenseful classics like *The Poseidon*

Adventure, *Earthquake*, and *The Towering Inferno*. These movies followed a standard formula. A wildly diverse ensemble cast separately perform individual deeds that, taken together, eventually contribute to salvation. On the bridge of the sinking ship, the fashion model gives her designer earrings to the wounded communications officer, and he uses them to jerry-rig the damaged radio. In the bowels of the ship, the Mexican cook gives his carrot peeler to the engineer, and he uses it to repair the water pumps. In the third-class cabins, a drunken ex-boxer uses brute force to open a bulkhead and rescue trapped passengers. There's no single narrative thread nor one single leader in charge, but their collective contributions cause them ultimately to prevail; the significance of all their individual efforts is apparent only afterward, when they are huddled together safely on shore, wet and injured but alive.

The story of the birth control pill follows the same formula.

The motley cast of characters responsible for the invention of the Pill sounds more like the cast of an Irwin Allen film than a product development team: a Swiss veterinarian, an eccentric chemist in the backwaters of Mexico, a discredited biologist, a septuagenarian feminist activist, a wealthy heiress, and a devout Catholic gynecologist. But this story commences not with a drug discovery team or a feminist campaign but with an esoteric group of professionals usually omitted from the annals of medical history—Swiss dairy farmers and their unusual method of supercharging the fertility of cows.

Dairy farms have always been big business in Switzerland. Consider the classic Swiss scene. Cows graze on high Alpine pastures with big brass bells jangling from their necks, an image as iconic as the Matterhorn. The goal of a dairy farm is to produce as much milk as possible, so dairy farmers must constantly impregnate their cows so that their udders will unceasingly produce milk. The dairy cycle is straightforward. A cow gives birth to a calf and after the calf is weaned the farmer immediately begins milking the cow's udders. Milk production is high at first, but after some months it declines. When the cow is finally "dried off"—exhausted of milk—the cow becomes fertile again, and the farmer hurries to mate the cow to begin the cycle anew. Calving, milking, mating, calving; such is the life of a Swiss dairy farmer—and a Swiss dairy cow.

This entire process, however, relies on one crucial skill. The farmer must be able to rapidly impregnate the cow after she is dried off. The most frustrating lament heard on any Swiss dairy farm is *Meine Kuh hat kein Kalb!*—"My cow has no calf!" A fertile cow is a cash cow, while an infertile cow is a money pit. The farmer must keep feeding the barren beast, but no milk is produced. In the late nineteenth century, the Swiss (a people frequently caricatured for their practical and unromantic nature) experimented with ways to accelerate a sterile cow's fertility. Eventually, one very practical and unromantic Swiss veterinarian discovered that if he inserted his hand into a cow's anus and crushed a fragile structure in the ovary by pinching it through the wall of the rectum, the cow rapidly became fertile again.

The intrepid vet's method soon became the "best practice" for the Swiss dairy industry, even though the farmers had no idea what exactly they were crushing. This peculiar proctological fertility technique remained unknown outside of the Swiss Alps until 1898, when a Zurich professor of veterinary medicine named Erwin Zschokke documented the procedure in a scientific journal for the first time. More importantly for the story of the Pill, Zschokke identified the precise anatomical entity that was being crushed, a yellowish ovoid structure in the ovary known as the corpus luteum.

In 1916, following up on Zschokke's discovery, two Viennese biologists showed that an extract taken from a female rat's corpus luteum suppressed ovulation, confirming that the structure had a role in inhibiting fertility. Later experimentation revealed that the active component of the extract was a steroid hormone called progesterone. These investigations were purely of academic interest, driven by curious scientists interested in deciphering the physiological intricacies of reproduction without any thought of the utility of their discoveries. No one imagined progesterone would have any practical use, least of all as an oral contraceptive for women. Yet Professor Zschokke's paper set in motion an elaborate and unrecognized pharmaceutical collaboration that stretched across nations, eras, and disciplines.

Based on the corpus luteum research, it was clear that progesterone played an essential role in the female reproductive system, and biologists around the world began to study the steroid hormone. However, their curiosity was tempered by one annoying fact. There was no method within synthetic chemistry for producing progesterone in a cost-effective way. The only known procedure produced tiny quantities of the hormone at a substantial cost. Demand for progesterone among biologists eager to study its effect on animal reproductive processes far exceeded the supply, rendering the hormone prohibitively expensive for most research. During the 1920s and 30s, one of the great unsolved puzzles in chemistry was figuring out how to (cheaply) synthesize progesterone—a puzzle that attracted the attention of a most unconventional chemist.

Kurt Vonnegut's science fiction novel *Cat's Cradle* tells the story of Nobel Prize–winning physicist Felix Hoenikker. Hoenikker is motivated by a pure curiosity unsullied by politics, greed, or common sense. One day, someone asks Hoenikker what kinds of games he enjoys playing during his downtime. Hoenikker replies, "Why should I bother with made-up games when there are so many real ones?" In the novel, the government hires the fictional Hoenikker to develop the atomic bomb as part of the Manhattan Project. He begins working on the bomb but abruptly stops. When the leaders running the Manhattan Project hurry into Hoenikker's lab to find out why, they discover the laboratory is crammed full of aquariums and turtles. Hoenikker has entirely switched his attention from the A-bomb puzzle to a tortoise puzzle: "When turtles pull in their heads, do their spines buckle or contract?"

Exasperated, the Manhattan Project leaders ask Hoenikker's daughter what to do. She tells them not to worry. The solution is simple. Her father merely works on whatever interesting thing he finds in front of him. Just remove all the turtles, she says, and replace them with atomic bomb research. They follow her advice and sure enough, when he returns to his laboratory the next day, not finding anything more interesting to play with, he resumes his work and eventually designs the first atomic bomb. There once lived a real-life version of Felix Hoenikker, and his name was Russell Marker.

As one prominent chemist observed, "There are more stories told about Russell Marker than perhaps any chemist. They are the campfire stories that bind our profession together." After a single year of graduate research in chemistry at the University of Maryland in 1925, Marker so impressed his advisor with his laboratory acumen that he announced Marker had accomplished enough to earn his doctoral degree. All Marker needed to do to graduate was complete a handful of courses to fulfill the university's standard graduation requirements. But Marker refused. Stunned, his adviser warned him that if he did not take the classes and get his diploma "he would end up as a urine analyst." Instead, the indifferent Marker dropped out of grad school. He took a research job at Ethyl Corporation, a chemical manufacturer specializing in hydrocarbons.

Ethyl was working on a problem that quickly seized Marker's attention. When comparing the effectiveness of different automobile engine designs, how can you distinguish between the contribution of the engine and the contribution of the gasoline? The reason this was such a mystery was because there are many different types of gasoline. Gasoline is not a single molecule or even a single consistent compound, but rather a highly variable mixture of thousands of possible hydrocarbon molecules. If a particular engine was performing badly, how could you tell if it was because of a poor engine design or because you were using a crappy gasoline?

In his very first year at Ethyl, Marker solved the problem. He developed a standardized system for grading gasoline that completely avoided trying to identify the mix of molecules in the gas and instead evaluated whether the gasoline exploded when you wanted it to explode. Marker's ingenious system compared the explosive behavior of a given batch of gasoline to a "perfect exploder" (iso-octane, which Russell assigned a rating of 100) and a "perfect non-exploder" (heptane, assigned a rating of 0). This became the octane rating system for gasoline that we still use at the gas pump today.

Despite his quick success at Ethyl, he soon became bored with hydrocarbon chemistry and after just two years at the company he resigned. Next, he took an academic research job at the prestigious Rockefeller Institute. Over the next six years, assisted by a single lab technician, Marker published an astonishing thirty-two papers on optical chemistry—a field completely different from hydrocarbon science—several of which are still considered classics in the field. But before long he wanted to play with new chemistry games, creating friction with his boss. "Levene just wanted me to keep doing the same old research on optical rotations," Marker explained later. "I was looking for something new." He switched jobs once again. This time he took a position as a chemistry research fellow at Penn State University, where he immersed himself within a new puzzle—the synthesis of progesterone.

Marker knew that this was one of the great unsolved chemistry riddles of his era, and in 1936 he set about trying to devise a technique that would create the steroid hormone on an industrial scale. His approach was completely different from what anyone else was trying, and brilliant in its simplicity. Steroids are very large molecules, a fact that makes them difficult to assemble. Synthesizing a large molecule is a game of addition: chemists start out with a small molecule, then methodically attach molecule after molecule, like building with Tinkertoys, until they have assembled the full steroid. But it is fairly easy to accidentally attach an intermediate molecule in the wrong place, ruining the entire synthesis and forcing you to start all over. Generally speaking, the larger the molecule, the more difficult the synthesis. Synthesizing a small molecule (like aspirin) is usually as easy as making mac and cheese. Synthesizing a large molecule (like progesterone) is more like preparing stuffed squab chaud-froid.

But Russell Marker turned the entire problem upside-down. Instead of viewing progesterone synthesis as a game of addition, he saw it as a game of subtraction. Rather than building the steroid out of smaller molecules, he decided to start out with an even *larger* molecule and whack off pieces until only progesterone was left. (In the jargon of chemistry, he intended to perform a degradation rather than a synthesis.) All he needed was a starting molecule that was even larger than progesterone.

Marker finally settled on a class of compounds known as phytosterols, large molecules similar to cholesterol but that are found in plants instead of animals. Marker attempted to prune away pieces of the phytosterol molecule in order to leave behind a progesterone molecule. He achieved success almost immediately by starting with a molecule of diosgenin—a type of phytosterol found in sarsaparilla roots—and degrading it into progesterone. It was a very good start, but even though he had demonstrated that his novel method worked, he still needed to show that his degradation technique could create progesterone on an industrial scale. To do that, he would need lots and lots of diosgenin. And that presented a new problem.

The stringy roots of the sarsaparilla plant were simply too meager a source of diosgenin to be useful in commercial production. Marker needed to find another plant with diosgenin—one that was large, cheap, and packed full of diosgenin molecules. He knew there were several species of plants containing diosgenin in the southwest United States, all tuberous plants with thick roots. So—like Valerius Cordus four centuries earlier—Marker embarked upon a plant-hunting expedition. In 1940, he ventured into the hot, unruly wildlands of Texas and Arizona. He tried root after root, but none of the thin American tubers produced enough diosgenin.

He eventually drifted south, crossing the Rio Grande into Mexico. There, in the state of Veracruz, Marker finally unearthed a plant with a remarkably high concentration of diosgenin. It was *Dioscorea composita,* the Mexican yam. These yellowish tubers boasted elephantine roots that weighed up to one hundred pounds, requiring a wheelbarrow to move them around. Marker hauled one fifty-pound yam back into the United States, bribing custom officials to allow him to transport the prohibited agricultural material across the border. Back at Penn State, he applied his degradation process to the diosgenin extracted from the yam. Success! The *D. composita* produced enough progesterone to enable industrial-level production.

Marker approached pharmaceutical companies and touted his clever degradation method, hoping to partner with one of them to produce commercial progesterone. These meetings did not go well. Marker was far more skilled as a chemist than as a pitchman, often lapsing into dull and highly technical disquisitions. But perhaps more damaging to Marker's cause, pharma execs were already skeptical about his unheard-of technique for creating the progesterone—and when they learned that this degradation process required gargantuan yams from a third-world country just a couple decades removed from a revolutionary war, they merely shook their heads in disbelief.

D. composita could grow only in the warm, dry climate of Mexico, which at the time was a disorganized and highly undeveloped country harboring dangerous anti-American sentiment against its wealthy and arrogant neighbor to the north. The pharmaceutical companies were convinced there was no way the yam could be reliably and safely collected in Mexico on the scale necessary for industrial production. Every pharma company that Marker approached turned him down.

Marker reacted to this setback in his usual manner. He resigned from his highly productive academic laboratory at Penn State and set up his own private lab in an old pottery shed in Mexico City. If the pharma companies would not collaborate with him, he would manufacture progesterone on his own. He paid Mexican laborers to unearth ten tons of yams, enough to fill a large truck, and began degrading their diosgenin. Working in isolation for two months, Marker produced an incredible three kilograms of progesterone—an amount that was quite probably more than the total supply of synthetic progesterone on Earth. Since progesterone was selling for $80 per gram—approximately $1,000 per gram in 2016 dollars—Marker had created the equivalent of three million dollars' worth of the hormone in his very first try. He had discovered a synthetic chemistry philosopher's stone: a means to transform yams into gold.

But he still needed a pharmaceutical business partner to help him distribute the hormone. He no longer wanted anything to do with the American pharma businesses that had turned up their noses at him. On the other hand, he knew nothing at all about the Mexican pharmaceutical industry and he only spoke a few rudimentary words of Spanish. Undaunted, he leafed through the Mexico City phone book until his finger came to rest upon a promising listing, a small pharmaceutical enterprise by the name of Laboratorios Hormona S. A.

The owners of Laboratorios Hormona were German and Hungarian Jews who had fled the rising anti-Semitism that consumed 1930s Europe. They joined forces with Marker and together founded a new company that they anointed Syntex S. A. The firm was devoted to manufacturing hormones using Marker's degradation process. But just as Marker's former bosses at Penn State, the Rockefeller Institute, and the Ethyl Corporation might have predicted, less than two years after Marker founded Syntex he sold all his shares in the company and left Mexico, abandoning all rights to Syntex's unprecedented supply of progesterone. He also abandoned science. He severed all ties with former friends and colleagues in chemistry and dropped out of sight in order to devote himself to a new passion: eighteenth-century silversmithing. This time, the interest stuck. Until the end of his days, Russell Marker spent most of his time fashioning intricate Rococo tureens and *surtout des tables*.

Like Vonnegut's fictional Hoenikker, Marker was never concerned with money or the practical utility of his research. He just liked to play with the "real games" of nature. Despite his eccentric and impractical nature, he had left behind what no other scientist had achieved: a breakthrough method for creating progesterone on an industrial scale.

Witnessing the out-of-nowhere success of Syntex, several American companies finally adopted the Marker degradation method, and by the early 1950s there were more than two hundred different progesterone compounds on the market. This abrupt glut of the hormone led to a flood of new research on female reproduction at academic laboratories around the world. One of these laboratories was in Cambridge, Massachusetts, and run by a Jewish biologist named Gregory Pincus.

In the early nineteenth century, a large influx of wealthy and educated German Jews immigrated to the United States and were quickly assimilated into American culture, becoming New York bankers, slave-owning plantation owners, western bordello madams, and Indian-fighting cavalrymen. The next wave of Jewish immigrants followed a very different path, however. Arriving at the end of the nineteenth century, these less well-to-do Eastern European Jews looked and sounded quite different from most Americans and mostly headed to inner city ghettoes like Manhattan's Lower East Side.

The old line Jews, already integrated into mainstream America, grew concerned about the new arrivals. Many of the established German Jews took it upon themselves to try to help Americanize their Eastern European brethren, and one of the most prominent examples of these charitable efforts was the Baron de Hirsch Fund. The Jewish philanthropist Maurice de Hirsch admired the way Norwegian immigrants in Minnesota had rapidly become good American wheat farmers. The perceived success of these Scandinavian émigrés inspired de Hirsch to put forth a straightforward idea. Instead of placing the incoming Eastern European Jews in ghettoes, what better way to convert them into full-blooded Americans than to turn them into farmers. Just as Norwegian immigrants became wheat farmers in Minnesota, de Hirsch's Fund would help Jewish immigrants become chicken farmers in New Jersey.

In 1891, the town of Woodbine, New Jersey was founded as an agricultural settlement for Eastern European Jews with the aid of de Hirsch money. The fund subsidized the purchase of farm land for Jewish immigrants and paid to train them for their new lives. However, de Hirsch's grand dream did not work out quite the way he envisioned. Most nineteenth-century European immigrants to the United States, including the Norwegians, had already been farmers in Europe. When they came to the New World, they brought with them extensive knowledge of farming. European Jews, in contrast, were mostly merchants and tradesmen. Instead of possessing farming skills, most of the Eastern European Jews arrived in the USA brought with them a long tradition of scholarly religious study, frequently analyzing religious texts for rabbinical guidance on everyday matters.

In Woodbine, the imported Jews applied their Talmudic skills to chicken farming. They contemplated the fowl and inquired, "*Vi tut a hun lebn*?" How does a chicken live? The Eastern European immigrants scrutinized the bird carefully, trying to unriddle how it produced eggs and how one might improve upon its egg-laying abilities. Since these Jews were used to setting up yeshivas (educational institutions) to study religious texts, it was quite natural for the Woodbine community to establish the Baron de Hirsch Agricultural College in 1894 to formalize their inquiry into the mysteries of the chicken. If these Jews were going to be farmers, they were going to be scholarly farmers.

Gregory Pincus was born in Woodbine in 1903, a member of the first generation of Woodbine-raised Jews. Two of his uncles were agricultural scientists at the Baron de Hirsch College, exposing him at an early age to the idea that it was possible to manipulate and improve upon the biology of Mother Nature. Studious and hard-working, Pincus received a Ph.D. in biology from Harvard

University, became an assistant professor in general physiology at Harvard, and then a professor of experimental biology at Clark University in Worcester, Massachusetts, where he founded the Worcester Foundation for Experimental Biology. In this academic laboratory, Pincus used progesterone to investigate what he called "the big questions": "Why does an egg start to develop and why does it continue to develop?"

Though Pincus seemed to have fulfilled de Hirsch's dream of integrating Eastern European Jews into the American way of life, Pincus remained an academic outsider. The period between the 1910s and the 1940s was the era of *numerus clausus,* a bigoted university quota system that restricted the number of Jews permitted at Ivy League institutions. Pincus looked different and spoke differently than his mostly WASPy colleagues. His foreign appearance would eventually contribute to a scandal that changed the course of his career.

At Clark, Pincus studied the eggs of the *Oryctolagus cuniculus,* a fluffy-tailed, buck-toothed laboratory rabbit. However, he soon found that it was difficult to precisely control the intricacies of rabbit fertilization. He began to wonder—instead of fertilizing the rabbit egg inside the rabbit—*in vivo,* in the vocabulary of biology—might it be possible to fertilize the egg *outside* of the rabbit? After several years of experimentation, he managed to fertilize a rabbit egg in a Petri dish. This was the very first *in vitro* fertilization of a mammalian egg.

Though Pincus did not seek any publicity for this achievement, the newspapers soon characterized Pincus as a modern day Frankenstein who was engineering "fatherless rabbits." Lending credibility to such charges was Pincus's personal appearance: his disheveled hair, crooked eyebrows, and dark and wild eyes made Pincus the spitting image of Rotwang, the deranged scientist who built a female robot in the contemporaneous film *Metropolis*. His notoriety soared when a reporter asked him if he intended to grow human beings inside of test tubes. Even though he actually replied, "I am *not* trying to create human life in the laboratory," the newspaper misprinted the quote as "I *am* trying to create human life in the laboratory."

For the rest of his career, Pincus was dogged by questions about his blasphemous (and nonexistent) process of "Pincogenesis." The fact that he was foreign-looking and Jewish only exacerbated the cloud of condemnation that hung over him. Though Pincus did his best to remain out of the public spotlight, the damage had already been done, and he found it difficult to raise funds for his research. He even took up an extra shift as a lab janitor to help keep his laboratory afloat. Discredited and isolated, Pincus was not sure how he would find enough financial support to enable his laboratory to resume its former luster investigating the big questions. Things looked fairly hopeless until he met Margaret Sanger.

Sanger was born in New York to a working-class Irish Catholic family—which in 1879 meant a very *large* family. Sanger believed that her mother, who had borne eleven children and endured seven miscarriages, perished at age fifty as a result of the debilitating effects of so many pregnancies. Staring across her mother's coffin, Sanger pointed her finger at her father and cried, "You are responsible for this! She had too many children!"

Sanger's hostility towards uncontrolled pregnancy was reinforced by her work as a nurse on the Lower East Side of Manhattan. One common feature in the lives of the poor immigrants to whom Sanger tended was the botched five-dollar back-alley abortion, pursued by desperate women who could not support another child. Sanger yearned for a cheap, convenient, and reliable method of birth control to help these women, but no new method had been developed since the invention of the diaphragm in 1842 for women and the full length condom in 1869 for men. In 1914, Sanger coined the term "birth control" and began to provide women with pamphlets and diaphragms—activities that violated federal law.

The anti-obscenity Comstock Act of 1873 made it illegal in the United States to disseminate information on contraception. In addition, thirty states had laws explicitly prohibiting the distribution of contraceptives. Consequently, during World War I, American servicemen were the only Allied troops not provided with condoms and, not surprisingly, American soldiers were afflicted with the highest incidence of sexually transmitted disease of all the warring countries.

Under the Comstock Act, Sanger was indicted in 1915 for sending diaphragms through the mail. She was arrested again in 1916 for opening the first birth control clinic in the country, in New York. But Sanger would not be thwarted. In 1921 she founded the American Birth Control League, the precursor to Planned Parenthood. For the following three decades she did everything in her power to raise awareness of birth control and to deliver contraceptives to American women. All the while she was obsessed with a singular dream. She envisioned a pill for women that could be taken like Aspirin that would enable them to take control of their pregnancies.

Sanger was no scientist. She did not know anything about the hormonal biology of reproduction, the science of pharmaceutical development, or even how the drug industry worked. She had no clue how feasible—or how wildly implausible—an aspirin-like birth control pill actually was. Though she repeatedly approached pharmaceutical companies with the idea of developing an oral contraceptive, they always rebuffed her, citing the Comstock Act and the fear of a Catholic boycott of all their products. "Besides," one pharma exec knowingly informed her, "Why would women want to take a pill every single day just to control conception?"

Despite her passion for a birth control pill, as 1951 rolled around, Sanger, now in her seventies, had all but given up. She had visited every major pharma company, some more than once, and had failed to persuade a single one of the potential value of such a drug; and she still had no idea about whether it was even scientifically possible to create the hypothetical pill. Sensing she was running out of time, she decided to switch tactics. Perhaps she could persuade a scientist to try to create a pill entirely on his own, outside of the pharma industry.

If she had possessed any knowledge of the realities of drug development in the 1950s, she would have realized just how unlikely it was for an academic scientist to create a novel medication at a university; post-FDA, the development costs for new drugs were prohibitive even for the best-funded academic laboratories. Oblivious to the extreme impracticality of her idea, she began to contemplate which scientist to target. He would have to be someone with a proven record of quality scientific research in female reproductive physiology. He would also have to be someone in a position so desperate they would be open to the overtures of a seventy-year-old feminist activist pursuing her wild dream of helping women control their fertility—and willing to pursue a controversial and probably illegal drug. Sanger eventually identified someone who checked off all the boxes. It was Gregory Pincus.

Though she had no ability to evaluate Pincus's scientific prowess, the same achievement that had publicly disgraced him and rendered him desperate—*in vitro* fertilization—persuaded her that he possessed the talent to create a birth control pill. Sanger invited Pincus to a dinner party hosted by the director of the Planned Parenthood Federation. By the end of the dinner, she had secured him a Planned Parenthood grant to support his existing research into animal fertilization. But she also outlined her real goal, the development of the world's first oral contraceptive. He confidently assured her that, yes, he could indeed develop such a drug—all he needed was a great sum of money.

Though the American businessman King C. Gillette is often considered the inventor of the disposable safety razor, it is more accurate to say that Gillette had the *inspiration* to create a disposable safety razor. He convinced a metallurgist named William Emery Nickerson to actually figure out how to turn Gillette's inspiration into a commercial reality. At the time, it was not known how to sharpen a thin square of steel to a razor-sharp edge, but with Gillette's financial support Nickerson managed to solve this tricky engineering problem. Margaret Sanger and Gregory Pincus formed a similar relationship. Sanger nursed a dream of an oral contraceptive, but had no idea how to convert her dream into reality. So she found someone who did. And just as Gillette funded Nickerson's research, Sanger found a way to fund Pincus. That way was Sanger's good friend Katharine Dexter McCormick.

McCormick's life started out like a storybook. She was born into an aristocratic Chicago family whose roots could be traced all the way back to the Mayflower. She studied biology in college and became the first woman to graduate from MIT. She married the dashing young Stanley McCormick, the heir to the massive International Harvester Company fortune. But before long, her charmed life crumbled around her. Her husband developed schizophrenia in his early twenties and was soon lost to incurable madness.

Believing that schizophrenia was hereditary, she vowed never to have children. So in the early 1900s, Katharine Dexter McCormick was a young, bright, beautiful woman with near-unlimited money, an insane husband, and no children. She needed something to occupy her keen mind, so she turned to one of the most prominent social causes of her time, the women's suffrage movement.

McCormick jumped into the fight for women's right to vote with "the force of a grenadier," as one friend described her, becoming the vice president of the League of Women Voters, funding the *Women's Journal,* and organizing much of the successful effort to gain ratification of the nineteenth amendment giving women the right to vote. During her suffragette days, in 1917, McCormick attended a lecture in Boston by a woman who immediately impressed McCormick with her passion and conviction. Margaret Sanger held great influence over McCormick from the moment she met her. And from the moment that Sanger first told McCormick about her dream of a contraceptive drug as easy to take as aspirin, the heiress was sold.

McCormick, educated as a biologist at MIT, believed in the power of biochemistry. After the successful ratification of the nineteenth amendment, Sanger's crusade for a birth control pill injected McCormick's life with a renewed sense of meaning and purpose. McCormick frequently aided Sanger's contraceptive efforts by smuggling diaphragms into the country for Sanger's birth control clinics. Yet, despite her immense wealth, McCormick was not able to fund research into a birth control pill. As her husband fell deeper and deeper into psychosis, she became ensnared in a litigious battle with her husband's family over his estate. McCormick was compelled to guide her philanthropy toward domains that her in-laws approved of, such as schizophrenia research.

Everything changed when her husband finally died in 1947. According to his generous will, she inherited full control over his $35 million estate—$350 million in today's dollars—making her, as one friend put it, "rich as Croesus." At long last, at the ripe age of seventy-two, Katharine Dexter McCormick was free to pursue her own agenda—and her agenda was oral contraception.

Sanger initially suggested that McCormick should fund research at multiple institutions around the world. But McCormick worried that such a scattershot approach would not work. She wanted a highly targeted approach that would produce a practical pill, not the open-ended uncertainty of basic research. After all, she was getting on in years and wanted to see the creation of a pill while she was still alive.

On June 8, 1953, Sanger took McCormick on a trip to Clark University in Massachusetts, where Pincus worked. He took the two septuagenarian ladies on a tour of his facilities—a short tour, since his lab was quite skimpy. Nevertheless, McCormick was persuaded by Sanger's enthusiasm and Pincus's confidence. "I believe you are the man to finally see our dream realized," McCormick said, and right there in the lab wrote Pincus a check for $40,000. This sizable sum (about $350,000 in 2016 dollars) was more than 1 percent of the entire National Science Foundation budget. Pincus, whose lab at Clark University had been struggling just to stay afloat, now had more funding than many of the top biology laboratories in the country.

This scandal-tainted Jewish outsider had just formed the unlikeliest of alliances with two elderly feminists, one fantastically rich, one raised in poverty, neither one at all qualified to judge the likelihood of his actually developing an effective oral contraception drug. Nevertheless, they all shared a common bond. They had all courted public controversy and faced public scorn. They were a battle-hardened group who recognized that they were embarking upon another war.

Pincus explained to Sanger that their goal was to develop an orally active version of progesterone. Ever since the earliest studies of the corpora luteum inspired by Professor Zschokke's paper on cow fertility, it was known that when progesterone was injected into a female mammal it inhibited ovulation. But progesterone did not work orally; the body would not take up the hormone through the digestive system. And even though it was possible in theory to convert an injectable version of a drug into an oral version, the oral absorption of drugs in animals' bodies differs from oral absorption in the human body. The only way to know for sure whether an oral version of progesterone would work was by testing it on people.

Through the 1960s, pharmaceutical companies usually did not even bother to manufacture a drug unless they already had a compound that could be swallowed, since devising an oral drug from an injectable one can be very expensive. When I was working for Squibb, we received FDA approval for an antibiotic called aztreonam, but the compound was active only via injection. We worked to put together a version we thought should be orally active, but to win FDA approval for the pill we would need to run expensive and time-consuming clinical trials—trials that might ultimately demonstrate the oral version did not work. Was there any way we could increase our confidence about the oral version before initiating the hassle and expense of clinical trials? There was. I swallowed the untested oral version of aztreonam myself.

Along with a few of my braver Squibb colleagues, one morning we all gulped down the drug with a cup of water, waited, then urinated into a cup for testing. Later that afternoon, the impromptu test results came back. Triumph! Our bodies had successfully absorbed the oral version of the antibiotic. That meant we could go forward with the clinical trials confident they would be worth the expense. However, as I celebrated at home that evening, I suddenly had to run to the bathroom to deal with a bout of wrenching diarrhea. Ironically, it did not even occur to me that my rogue testing could have caused my gastrointestinal distress. I wanted the drug to work so badly that I never considered the possibility that my incessant bathroom visits were due to the oral aztreonam. Instead, I recalled that I had eaten egg salad for lunch—and convinced myself that my waves of diarrhea must have been the result of food poisoning from spoiled egg salad. I forgot about the entire incident until the clinical trials commenced and multiple subjects reported explosive diarrhea. Needless to say, we never received FDA approval of our orally active drug.

Pincus started searching for an oral version of progesterone by testing progesterone compounds on rabbits. There were more than two hundred different commercially available progesterone compounds—all produced using the Marker degradation method—and Pincus fed every one of them to the rabbits in his Clark University laboratory. Three of the compounds reliably prevented pregnancy without producing adverse effects. That was enough. Now he could test these three drug candidates on humans.

There was one final hurdle to surmount, and it was a big one. According to federal law, only a clinical physician could direct human drug trials. Pincus would need to find a partner who was willing to endure the inevitably scrutiny and controversies that would attend such a scandalous project—a project that, strictly speaking, violated both state and federal anti-contraception laws. Pincus must have wondered if it was easier to get a $40,000 check than to locate a physician willing to test the world's first oral contraception.

A silver crucifix always graced the wall of Dr. John Rock's office. The lifelong Catholic attended mass at seven in the morning every day, at St. Mary's in Brookline or sometimes at the Immaculate Conception Church. He was always gracious and unwaveringly polite, holding the door for his patients at Harvard Medical School and always addressing them as Mrs. or Miss. Rock taught obstetrics for more than three decades at Harvard, and he believed that the most prominent form of suffering among his patients was the anguish of unwanted pregnancies.

He had seen destroyed wombs, premature aging, and financial catastrophe all resulting from a mother having too many babies. Although Rock was a staunch social conservative—early in his career he opposed the admission of women to Harvard for their own good—he had slowly developed progressive ideas about birth control. Despite the Catholic Church's strident opposition to contraception, Rock believed that birth control could reduce poverty and eliminate the medical problems resulting from repeated pregnancies. He felt certain that Christ would approve of birth control.

In the 1940s, Rock began teaching his Harvard students about contraception, something unheard of in medical schools at the time. He believed that if people merely heard the logic and facts, they would come to embrace birth control as both rational and compassionate. He published a book about birth control, which he thought would lead to a sea change in people's attitudes. It did not. However, it did draw the attention of a Jewish biologist at Clark University.

After completing his progesterone trials on rabbits, Pincus bumped into Rock, an old acquaintance from his Harvard days, at a medical conference. Knowing about Rock's progressive position on birth control from his book, Pincus struck up a conversation about the possible use of an oral progesterone as a contraceptive, hoping to feel out whether Rock might be interested in running human trials with him. To Pincus' utter astonishment, Rock informed him that he was already testing progesterone on his patients—on *infertile* women.

While Pincus had been giving progesterone to rabbits to directly inhibit their ovulation, Rock was giving it to women to indirectly stimulate their ovulation. Rock's methodology struck Pincus as counterintuitive. The gynecologist injected his patients with a daily regimen of progesterone for several months, theorizing the inhibitory effects of the drug would allow their bodies to rest from the "stress" of ovulation. Then, after ceasing the progesterone injections, Rock speculated that the women's well-rested reproductive organs would "rebound" vigorously, enabling them to conceive more easily. Remarkably, Rock's instincts proved correct.

After Rock administered his daily progesterone therapy to eighty women, thirteen of them conceived within four months of finishing the hormone treatments—an astonishing number in fertility research at the time. This effect became known as the "Rock Rebound." But for Pincus, the real headline from Rock's research was the fact that Rock was already testing progesterone on human beings.

Even so, Rock was sixty-eight—an age when most physicians began to settle into a comfortable and noncontroversial retirement. Pincus suspected that Rock might be hesitant to participate in a project as scandalous and demanding as human trials of an oral contraceptive. But to Pincus's delight, Rock readily agreed to join the project.

Pincus felt that Rock was the perfect choice to oversee the clinical trials. Still stinging from the negative publicity that had dogged his in vitro fertilization research, Pincus hoped Rock's prestige, handsome looks, and staunch Catholic faith would help deflect the inevitable blowback once their contraception research became publicly known. For his part, Rock was confident—many would have said naively confident—that the pope would approve of a progesterone-based oral contraceptive; after all, progesterone was a natural hormone that already existed in the body in order to prevent fertility and therefore should be an acceptable form of birth control. Surely the pope would appreciate the Christian need to help poor women control their pregnancies.

Pincus was not merely concerned with negative publicity. The Comstock Act was still in full force, as were Massachusetts's rigid anti-birth control laws banning the distribution of contraceptives. Together, Rock and Pincus devised a way around the anti-contraception laws. Taking advantage of Rock's existing studies, they would conduct the oral progesterone trials as "fertility research" rather than "contraception research." Although Pincus and Rock camouflaged its true purpose, the study would be historic, nothing less than the first human trials of an oral contraceptive.

In 1954, Rock gathered together fifty female volunteers from his fertility laboratory and began giving them the three versions of the progesterone that Pincus had successfully tested on rabbits. Month after month, Rock carefully checked whether these women were ovulating. Not one ovulated while taking oral progesterone. Simultaneously, in a decision that would be considered highly unethical by modern standards (though typical for the time), another group of patients was given the progesterone compounds without their consent. Twelve female and sixteen male patients at the Worcester State Mental Hospital were used as guinea pigs to evaluate the rudimentary safety of the drugs and see whether they produced any adverse physiological effects. Fortunately for the twenty-eight psychiatric patients, they did not.

Pincus and Rock were ecstatic. But there was still one more crucial question to be answered. Even though the progesterone pills did not seem to elicit any obvious physical problems, Pincus and Rock fretted that the hormone might damage women's reproductive systems. Specifically, would women become fertile again after they stopped taking progesterone? Yes, they did. The oral contraceptive was not only effective but temporary, eliminating any fear that the Pill would have a permanent sterilizing effect.

After the success of the Boston trials, Rock and Pincus were confident they had a true oral contraceptive in hand. Pincus and Rock selected one of the three progesterone compounds known as noretynodrel for all future drug development efforts, based on results from animal studies that suggested that it had the least potential for side effects. But to actually turn noretynodrel into a commercial product, they needed FDA approval. And to get FDA approval, they needed to run more comprehensive experiments on humans. But since clinical trials of a contraceptive were both prohibited by law and contrary to religious dogma, how could Pincus and Rock get the necessary trials off the ground?

Hoping to find a location outside the reach of the law, Pincus visited the island of Puerto Rico in the summer of 1951. It was perfect. The American territory was densely populated and one of the poorest regions in North America, conditions that made Puerto Rican officials very supportive of birth control methods. At the time, many American companies were building factories in Puerto Rico and women could find good-paying jobs . . . if they could find a way to control their pregnancies. Even better, sixty-seven separate clinics across the island were already sharing non-drug-based methods of contraception with women.

In April, 1956, Pincus and Rock initiated their first drug trial at the clinic in the town of Rio Piedras. As soon as word got out that the clinic was offering a drug to prevent pregnancy, the trial filled to capacity with female volunteers. Encouraged, Pincus and Rock quickly expanded the trials to other clinics. After a year of testing, the results came in. Pincus and Rock were joyous. The Pill was 100 percent effective when taken properly.

Nevertheless, there was a major caveat to these wondrous findings. About 17 percent of the women in the study complained of nausea, dizziness, headaches, stomach pain, or vomiting. In fact, the director of the Puerto Rico trials informed Pincus that a 10-milligram dose of progesterone produced "too many side reactions to be generally acceptable." Rock and Pincus brushed aside the warning. With an attitude distressingly common among drug hunters who believe they are *this* close to success, the two men suggested that the women's complaints might be psychosomatic. After all, their patients in Boston—whom Rock had inspected in person—experienced far fewer negative reactions. The two male drug hunters dismissed nausea and female bloating as minor nuisances compared to the outstanding benefits of their new medication.

This disgraced biologist and the ivory tower idealist, working without the support of either industry or academia, having dodged federal and state laws by holding their drug trials in an offshore American territory and willfully ignoring disturbing signs of deleterious side effects, had nevertheless demonstrated that it was possible to create an inexpensive and reliable oral contraceptive. Now all they needed was a way to manufacture and distribute the possibly unsafe drug on an industrial scale so it could be provided to any woman who needed it. Of course, there was only one kind of institution capable of making and selling drugs on an industrial scale. Big Pharma.

When Pincus had first approached the pharmaceutical company G. D. Searle in the early 1950s about funding research for an oral contraceptive, Searle's response was a resounding no. At the time, many pharmaceutical companies were raking in profits from several new kinds of miracle drugs, including antibiotics, psychiatric drugs, and glucocorticoids—a recently discovered class of medication, such as hydrocortisone, that possessed amazing anti-inflammatory properties. Since glucocorticoids were used to treat everything from poison ivy to autoimmune disease, they were flying off the shelves. Searle had built up a very lucrative glucocorticoid business, so the idea of making a controversial drug that might put their other sales at risk of a Catholic boycott was a non-starter. Searle executives believed that such a boycott could result in the loss of one fourth of Searle's employees and a considerable portion of its hospital business.

Moreover, beyond the legal and religious risk factors, Searle executives simply did not believe there would be much of a market for an oral contraceptive. The prevailing wisdom among the all-male executives was that no healthy woman would ever willingly take a drug that neither treated nor prevented disease—especially a drug they would need to take every single day. However, when Pincus and Rock returned from Puerto Rico with the results from their surreptitious trials, Searle completely reversed its long-standing position.

Though Pincus and Rock believed it was their hard-earned data that did most of the convincing, Searle was secretly influenced by developments of which Pincus and Rock were unaware. Searle had already marketed progesterone to women as a treatment for various gynecological disorders. To the amazement of Searle executives, many of these women spontaneously began to use the drug as an ad hoc contraceptive method, a use that Searle had in no way encouraged and that was definitely not approved of by the FDA. Thus, Searle was already open to the possibility of a market for an oral contraceptive when Pincus and Rock showed up on their doorstep with FDA-ready human data.

In a historic decision, Searle agreed to move forward with the manufacture of the first commercial oral contraceptive. Fortunately, they did not overlook the troubling side effects from the Puerto

Rico trials, which the pharma company took very seriously. Searle scientists adjusted the formulation of Rock and Pincus's synthetic progesterone compound to reduce the amount of breakthrough bleeding and other adverse symptoms. The result was a small white tablet, not terribly different in size and weight from an aspirin. Sanger was ecstatic. The impossible dream of a lifelong feminist had somehow become reality.

Searle anointed the pill with the trade name Enovid. The FDA approved Enovid as a contraceptive in February 1961, and five months later Searle started marketing Enovid to the public, seven years after Gregory Pincus had received his first check from Katharine McCormick and fourteen years after Russell Marker had set up his private progesterone lab in a Mexican pottery shed. At the age of eighty-five, Katharine Dexter McCormick celebrated the event by being one of the very first women in the United States to walk into a drugstore to have her birth control pill prescription filled.

Within two years of Enovid's release, 1.2 million American women were on the Pill. By 1965, this number had risen to five million. The drug that no company wanted to touch turned out to be Searle's best-selling product for more than a decade, far outstripping its sales of glucocorticoids. By the late 1960s, seven pharmaceutical companies were producing oral contraceptives and more than twelve million women were taking the Pill worldwide. Today, more than 150 million prescriptions for the Pill are written each year.

There are few medical inventions in history that transformed the basic fabric of society so quickly and so dramatically. Rock and Sanger both viewed the pill primarily as a public health measure, to prevent physical deterioration from excessive pregnancies, and secondarily as a way to improve the financial stability of impoverished women who could not afford to raise additional children. Arrayed against them were social conservatives who argued that the Pill would encourage women to engage in society-destroying promiscuous sex. But the reality was quite different than what anyone imagined.

"'Someone said once that not everyone with vocal cords is an opera singer. And not everyone with a womb needs to be a mother," asserted Gloria Steinem. "When the Pill came along we were able to give birth—to ourselves." Women could now pursue a career as a doctor, lawyer, or business executive on their own timetable. The average family size plummeted, and family size soon became inversely proportional to family income, a clear indication that birth control was fully embraced by the educated and wealthy classes.

The Pill made it possible for women to control their fertility without depending on their partners, and in a manner that was disconnected from the sexual act itself. While the Pill was certainly not the first contraceptive *intended* to work this way—the sixth-century medical writings of Aetios of Amida advised women to avoid pregnancy by wearing cat testicles in a tube tied around the waist—the Pill was the first one that actually worked.

Lynne Luciano, a history professor at California State University with an interest in women's issues, points out how the Pill changed society's basic perception of sex. "In psychology journals, prior to 1970, frigidity was listed as a major problem for women. Today, frigidity has practically vanished from the literature. It's been replaced by erectile dysfunction and premature ejaculation, which were never considered problems before."

Not everything changed, however. Ever the idealist, John Rock had always maintained that oral contraception was compatible with the Catholic faith. The pope thought otherwise and explicitly banned the Pill in the encyclical "Humanae Vitae," a policy statement authored by Pope Paul VI in 1968 to reaffirm the orthodox teachings of the Catholic Church. Yet when Rock was confronted with the Church's opposition to his revolutionary drug, instead of terminating his involvement with oral conception, he discovered that he was an idealist first and a Catholic second. After a lifetime of attending daily mass, he ceased going to church altogether. Despite the pope's ban, millions of Catholic women around the world also chose to follow their own conscience and committed the sin of swallowing the little white tablet.

The Pill did not originate in a Big Pharma science lab or a sales team meeting. First, Swiss dairy farmers who wanted to make their cows pregnant faster made a peculiar anatomical discovery. Then, a veterinary professor published this finding, leading to the identification of progesterone as an anti-ovulation drug. An eccentric and solitary chemist figured out how to make progesterone simply because it was an interesting puzzle. Two septuagenarian feminists selected a discredited biologist to realize their dream of creating an oral contraceptive. A devout and hopelessly idealistic Catholic gynecologist agreed to run the world's first human trials of the oral contraceptive. Together, the biologist and gynecologist dodged federal and state laws—and medical ethics—by holding trials in Puerto Rico and ignoring clear signs of adverse side effects. They only succeeded in convincing a pharmaceutical company terrified of Catholic boycotts to manufacture the drug after the company fortuitously noticed that women were spontaneously using one of their other drugs for the off-label purpose of contraception.

This, in a nutshell, is why it is so hard to develop new medicines. Imagine you want to replicate this process: "Can we develop a cure for baldness the same way we developed a birth control pill?" To become a successful drug hunter requires talent, moxie, persistence, luck—and even then, it might not be enough. And we should not overlook Big Pharma's frustrating and unhelpful role in this process. Every single pharma company rejected Pincus and Sanger's proposals when they solicited the companies for help developing the Pill. A previously hostile pharma company jumped in only after an independent team of drug hunters sweated and bled their way to an FDA-approvable clinical trial entirely on their own.

The modern drug development process is drastically unfair and completely unreasonable, and yet it still managed to significantly improve the lives of hundreds of millions of women. And this is the true nature of drug hunting.

12 Mystery Cures

Discovering Drugs Through Blind Luck

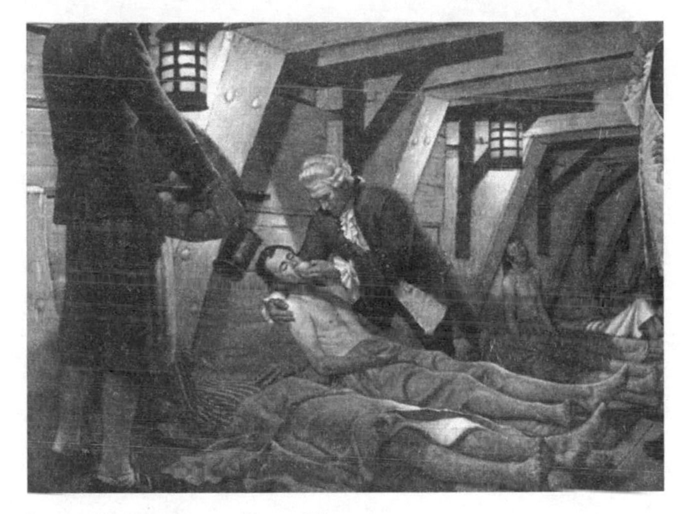

James Lind treating sailors with scurvy

"A sick thought can devour the body's flesh more than fever or consumption."

—Guy de Maupassant, *Le Horla et autres contes fantastiques*

One of the most basic truths of drug hunting is the uncomfortable fact that the vast majority of important drugs were discovered without the foggiest idea of how the drug actually worked. It often takes decades before researchers decipher how a new drug fully operates on the body. In many cases, despite generations of investigation, we still do not fully comprehend how a particular medication works. For instance, as of 2016, gaseous surgical anesthetics (such as halothane), modafinil (a narcolepsy drug), and riluzole (an ALS drug) all remain pharmaceutical mysteries. For physicians, this lack of understanding can be somewhat unsettling. But for the drug hunter, it can be liberating.

Anyone with an alert mind stands a chance of identifying a potentially useful compound and converting it into a medication, even if they possess little knowledge of biological mechanisms. During the Age of Plants, of course, drug hunters possessed zero understanding of how medicines worked. Drug discovery was 100 percent trial and error. Until Ehrlich proposed his receptor theory in the early twentieth century, theories of how drugs worked ranged from the misguided (such as the proposition that drugs changed the shape of cells) to the ludicrous (such as the conviction that the cure for a given disease would come from a plant that physically resembled the diseased organ). Even so, sometimes even the most ignorant of beliefs can serve as the catalyst for a crucial discovery. Simply having the motivation to proceed—any motivation—can stir a drug hunter to continue down the rugged path of exploration. In fact, the very first scientific experiment on the curative effects of a drug was the result of a fallacious assumption.

Scurvy is a horrible affliction recognized since antiquity. In the fifth century BC, Hippocrates identified its symptoms of bleeding gums and body-wide hemorrhaging, followed by death. Scurvy was fairly uncommon in ancient times, however, because ocean voyages were rarely very long. But the disease began to explode at the dawn of the fifteenth century as Europeans began to tackle extended sea journeys as they ventured to distant continents. In the midst of a sustained ocean cruise, vigorous and healthy sailors would suddenly collapse.

Some historians state that scurvy caused more deaths in the British fleet in the eighteenth century than from all French and Spanish arms combined. Richard Walter, the chaplain of Commodore George Anson's failed attempt to circumnavigate the globe, wrote up an official account of the voyage. Anson set out from England on September 18, 1740, with six warships and 1,854 men. By the time the expedition returned home four years later, only 188 remained alive. Most had perished from scurvy, which Walter documented in his reports. He described ulcers, difficult respiration, rictus of the limbs, skin as black as ink, teeth falling out, and—perhaps most disquieting of all—a foul corruption of the gums that endowed the victim's breath with an abominable odor.

Scurvy also seems to affect the nervous system by shutting down sensory inhibitors, making the victim extremely sensitive to taste, smell, and sounds. The fragrance of flowers on the shore can cause a victim to moan in agony, while the crack of gunfire can be enough to kill a man in the advanced throes of the disease. In addition, victims' emotions often become unmanageable, so that they cry out at the slightest disappointment and yearn disconsolately for home.

In the eighteenth century, no one knew what caused scurvy, so no one had any idea how to prevent it or treat it. The medical establishment's best guess was that scurvy was a disease of putrefaction and thus best treated by acids, including elixir of vitriol (sulfuric acid), preparations believed to slow the rotting process. It was not clear that the acid treatments were helpful, though, so eventually a Scottish physician decided to put the acid theory to the test.

James Lind was appointed the ship's surgeon of the HMS *Salisbury* in the Channel Fleet in 1747. After the ship was two months at sea, sailors began falling ill with scurvy. Lind took the opportunity to launch his experiment. His approach was sensible and straightforward: he applied a variety of different acids to his scurvy patients and evaluated the results. Lind divided twelve of the sick sailors into six groups of two, an extremely small sample size by modern standards. All his patients received the same diet, but each pair of sailors was treated with a different kind of acid. The first group received a quart of (lightly acidic) cider, the second group received twenty-five drops of elixir of vitriol (the most highly regarded remedy at the time), the third group received six spoonfuls of (lightly acidic) vinegar, the fourth group received two oranges and one lemon because citrus fruits are acidic, the fifth group received a spicy paste plus a drink of barley water (spices were another common treatment of scurvy because their effects were believed to be similar to acid). The sixth group, meanwhile, received a half a pint of seawater; this placebo treatment made the final pair of sailors the very first control group in a clinical drug trial.

After six days Lind ran out of fruit, so he had to terminate his tests on group four. Yet, amazingly, one of the citrus-treated sailors was already fit for duty while the other had almost completely recovered. None of the other sailors had recovered at all except for the pair treated with cider, who showed a mild improvement. Today, of course, the interpretation of these results is obvious. We now know that scurvy is a disease caused by a dietary deficiency of vitamin C, a compound required for the synthesis of collagen. Collagen provides the strength, structure, and resiliency for our connective tissues, including our blood vessels, and without enough collagen our connective tissue breaks down and produces the symptoms of scurvy, including bleeding and the reopening of old wounds. Citrus fruits contain high levels of vitamin C, while apple cider contains small amounts of vitamin C; none of the other treatments that Lind employed contain any vitamin C. Since fruits and vegetables could not be stored on long sea voyages, eighteenth-century sailors subsisted on cured meats and dried grains—a diet that lacked vitamin C.

Vitamin C itself was not discovered until the 1930s, almost two centuries after Lind's pioneering experiment. So when Lind published *A treatise of the scurvy* in 1753, sharing the results of his acid evaluation, his findings was largely ignored. Even though he had shown that citrus fruits and apple cider were effective treatments for scurvy, he had no idea why, and without the why most physicians still clung to their familiar (but useless) acid treatments. Over time, though, many officers and surgeons came to realize that Lind was correct and that citrus fruits were indeed an effective answer to scurvy. More and more ships began providing their sailors with citrus fruits and citrus drinks on long journeys, dramatically reducing the incidence of the gum-rotting disease. Finally, in 1795—four decades after Lind's study—the British navy officially adopted lemons and limes as standard issue at sea. It took almost another decade before the British naval supply chain could provide adequate citrus to their ships all around the globe. Limes came to be most popular since they were abundant in the British West Indian colonies (unlike lemons), leading Americans to endow British sailors with the nickname "limeys."

One reason it was so difficult to identify the active ingredient in citrus fruits that prevented scurvy was that scientists were not able to produce scurvy in animals. Eventually, the medical establishment came to believe that scurvy was a disease that afflicted only *Homo sapiens.* Since it was not possible to conduct scurvy experiments on animals, the only way to test out the effect of different citrus fruit compounds was to use scurvy-rotting humans—but who would be willing to volunteer to endure the disgusting, painful disease for a medical experiment, especially considering that you might not even get treated with an effective compound? As a result, there was little progress in understanding how citrus fruits worked, until a stroke of good fortune befell two Norwegian scientists in 1907.

Alex Holst and Theodor Frolich were trying to induce beriberi in animals, a disease now known to be caused by a lack of vitamin B1. They fed guinea pigs a diet limited to grains and flour hoping to produce beriberi. To their surprise, the guinea pigs developed scurvy instead. This was a wildly lucky turn of events, because virtually every species of mammal is able to synthesize its own vitamin C within its body and thus does not require the vitamin in its diet. Holst and Frolich had fortuitously stumbled upon one of the precious few species other than humans that do not produce vitamin C internally. The two scientists recognized that they had just come up with an animal model of scurvy. Several teams began trying to identify the active ingredient in citrus fruits that prevented scurvy, and in 1931 scientists finally identified L-hexuronic acid as the critical compound. It was later renamed ascorbic acid, from *a-* ("no") and *-scorbutus* ("scurvy"). It took another twenty-five years before scientists identified ascorbic acid's role in building collagen. Thus, it was more than two centuries from the time that James Lind identified an effective anti-scurvy drug until the medical establishment finally unraveled how the drug worked.

Perhaps the broadest and most frequently prescribed class of "mystery" drugs today is the psychoactives—medications for mental illness. All the way into the 1950s, not only was there no therapeutic drug for schizophrenia, depression, or bipolar disorder, most members of the psychiatric establishment believed there could *never* be a drug to treat these disorders, since it was widely believed that mental illness was primarily due to unresolved childhood experiences. This was the central conviction of Sigmund Freud, whose theory of mental illness—known as psychoanalysis—swept through the United States in the early twentieth century. (Ironically, Freudianism was almost completely wiped out in Europe for the same exact reason it became so popular in America. The vast majority of the early psychoanalysts were Jews, as was Freud himself, and as the Nazis rose to power in Hitler's Germany, these Jewish psychoanalysts fled Europe for the safety of American shores, moving the world capital of psychoanalysis from Vienna, Austria, to New York City. It was as if the Holy See of the Catholic Church moved from the Vatican to Manhattan.)

By 1940, psychoanalysts had taken over every position of power in American psychiatry, controlling university psychiatry departments and hospitals and completing a hostile takeover of the American Psychiatric Association. In addition, psychoanalysts drove a profound change in the basic nature of American psychiatry. Before the Freudians fled Nazi Europe, American psychiatry consisted almost entirely of "alienists"—psychiatrists who tended to the severely disabled mentally ill within mental institutions far away from population centers; the fact that mental asylums stood apart from good society gave rise to the moniker "alienist." But the Freudians brought psychiatry into the American mainstream by insisting that everyone was "a little mentally ill" and that they could be fixed through relaxing therapy sessions in the comfortable offices of the psychoanalyst. Thus, the Freudians moved psychiatry out of remote, isolated institutions onto the couches of downtown offices and suburban homes.

Since psychoanalysts believed that patients could be cured only through "talk therapy"—exploring their childhood experiences through dreams, free association, and frank confession—they were convinced that no chemical could possibly bring about any positive change in a person suffering from mental illness. Consequently, there was absolutely no support for drug hunters questing for psychiatric medications. Through the 1950s, no Big Pharma company pursued any kind of program for mental illness drugs, no academic laboratory was looking for mental illness drugs, and very few mainstream hospitals were looking for evidence that a drug might improve the condition of its mental patients. Though there were still a few non-Freudian alienists who were dealing with severely sick schizophrenic and suicidal patients in remote mental asylums who still held out hope that there might one day be a pharmaceutical remedy, the entire medical profession took it for granted that there would never be a Salvarsan or insulin for mental illness. In such a hopelessly anti-drug environment, the only real hope for the development of a psychiatric drug was a false hypothesis and blind luck. But false hypotheses and blind luck have always been key ingredients of successful drug hunting.

Henri Laborit was not a psychiatrist and, indeed, knew very little about psychiatry at all. He was a surgeon in the French Navy who served in the Mediterranean squadron during the Second World War. During the war, he became interested in finding new medications to aid in surgery, hypothesizing that a drug that induced artificial hibernation in patients would lessen the dangers of shock following surgery. Pursuing this line of thinking, Laborit speculated that any drug that lowered patients' temperatures might help induce artificial hibernation.

Working in a French military hospital in Tunisia, Laborit obtained a new kind of antihistamine compound from a colleague that was believed to lower body temperature, a compound known as chlorpromazine. He tried chlorpromazine out on his surgery patients, hoping to reduce the severity of post-surgical shock. But Laborit noticed that even before he had a chance to administer an anesthetic, the patients' attitude underwent a dramatic mental change. The chlorpromazine made them indifferent toward the major surgery they were about to undergo, an indifference that continued after the surgery was completed. Laborit wrote about this discovery, "I asked an army psychiatrist to watch me operate on some of my tense, anxious Mediterranean-type patients. After surgery, he agreed with me that the patients were remarkably calm and relaxed."

Chlorpromazine, it turned out, did not induce any kind of artificial hibernation and, indeed, has little effect on body temperature. But Laborit was impressed by the unexpected psychological effects of the drug. He began to wonder if the compound could be used to help mitigate psychiatric disturbances. In 1951 Laborit returned to

France, where he persuaded a healthy psychiatrist to take an intravenous dose of chlorpromazine in order to describe the subjective effects of the drug. The psychiatrist guinea pig initially reported "no effects worthy of mention, save a certain sensation of indifference." Then, he abruptly fainted. (Chlorpromazine also has anti-hypertensive effects, reducing blood pressure.) After that, the head of psychiatry at the hospital forbade further experimentation with chlorpromazine.

Undeterred, Laborit moved to another hospital where he tried to convince the psychiatrists to administer the drug to their psychotic patients. The physicians refused, which was not very surprising considering that most psychiatrists believed that the only way to control (though not treat) schizophrenics was through the use of powerful sedatives—and chlorpromazine was no sedative. But Laborit did not give up. He finally persuaded a psychiatrist to run a test using his "indifference" drug.

On January 19, 1952, the psychiatrist administered chlorpromazine intravenously to a patient known as Jacques L., a twenty-four-year-old psychotic who was highly agitated and violent. Jacques rapidly settled down and maintained a state of calm for several hours. And then, a miracle. After three weeks of receiving daily doses of the drug, Jacques was able carry out all his normal activities. He could even play an entire game of bridge without any disruptions—something that was previously unthinkable. He recovered so well, in fact, that the astonished physicians discharged him from the hospital. The psychiatrists had just witnessed something completely unheard of in the annals of medicine: a drug had almost entirely eliminated the symptoms of psychosis, enabling a previously uncontrollably violent patient to return to the community.

Chlorpromazine was marketed to the public by the French pharmaceutical company Rhône-Poulenc in 1952 under the trade name Largactil. The following year it was offered in the United States by Smith, Kline, and French under the trade name Thorazine. It bombed. Nobody prescribed it. Most psychiatrists did not think it was possible even in principle for a drug to treat the symptoms of mental illness. American shrinks dismissed chlorpromazine as a distraction that concealed rather than cured the true childhood sources of a patient's illness, with several prominent psychiatrists deriding Laborit's drug as "psychiatric aspirin."

Smith, Kline, and French was stunned. They were offering for sale the first miracle drug proven to treat the symptoms of psychosis, yet psychiatrists refused to use it. The pharmaceutical company finally hit upon a solution. Instead of trying to coax psychiatrists into prescribing the drug, Smith, Kline, and French salesmen targeted state governments by arguing that if state-funded mental institutions used chlorpromazine, they would be able to discharge patients instead of warehousing them forever, drastically cutting costs and reducing the state's bill. A few of these state institutions—more concerned with their bottom line than abstruse debates about the philosophy of mental illness—gave chlorpromazine a try. All but the most hopeless patients exhibited dramatic improvements, and just as Smith, Kline, and French had promised, many were discharged back into society.

Smith, Kline, and French's revenues increased eight-fold over the next fifteen years. By 1964, more than 50 million people around the world had taken the drug, which quickly became established as the first line of treatment for any schizophrenic patient. Individuals who were once lost for life in the quasi-dungeons of public asylums could return home and, amazingly, live engaged and productive lives. The success of chlorpromazine also marked the beginning of the end for psychoanalysis and the Freudian dominance of American psychiatry. After all, why would you want to spend week after week for years on end sitting on a shrink's couch talking about your mother when you could swallow a pill instead—and watch your symptoms disappear?

All the antipsychotic drugs we use today, including olanzapine (Zyprexa), risperidone (Risperdol), and clozapine (Clozaril), are chemical variants of chlorpromazine. In the more than sixty years since the clinical adoption of chlorpromazine, the scientific community has not been able to come up with a fundamentally better approach. And yet, we still do not have a clear idea of exactly how chlorpromazine mitigates the symptoms of schizophrenia. But that never stopped every pharmaceutical company from trying to make a chlorpromazine knockoff.

Other drug makers wanted to duplicate the blockbuster success that Rhône-Poulenc and Smith, Kline, and French enjoyed with the world's first antipsychotic. So they assembled their own teams to try to synthesize their own variation of the chlorpromazine compound. One of these hopeful copycats was the Swiss pharmaceutical company Geigy, a corporate ancestor of Novartis. Geigy executives reached out to Roland Kuhn, a Swiss professor of psychiatry who had a strong interest in finding new treatments for mental illness. Geigy supplied him with a chlorpromazine-like compound that the company had labeled G 22150 and asked him to try it out on his psychotic patients. The drug produced extreme, intolerable side effects, rendering it unfit as a treatment. So in 1954, Kuhn asked Geigy for a new compound to try.

Kuhn met with Geigy's head of pharmacology at a hotel in Zurich where Kuhn was presented with a large chart filled with forty hand-scribbled chemical structures. The Geigy executive asked Kuhn to pick one. He pointed to the compound that appeared to be the most similar to chlorpromazine, a compound labeled G 22355. It would prove to be a very fateful choice.

Kuhn returned to his hospital and administered G 22355 to a few dozen psychotic patients. Not much happened; certainly not the dramatic reduction of symptoms elicited by chlorpromazine. You might expect that Kuhn would have once again returned to Geigy and selected another compound from Geigy's chart of chemicals. Instead, Kuhn decided to try something else. Without informing Geigy, he decided to give G 22355 to some of his patients suffering from depression.

It had been just a few years earlier, as we saw, that the first antipsychotic had been discovered—not because of any Big Pharma research project but accidentally, by a surgeon in Tunisia who was trying to reduce surgical shock. And now in Switzerland a psychiatrist decided to ignore the task that he had been hired to do—finding a new antipsychotic—and instead decided to test a failed antipsychotic drug on patients suffering from depression. Why? Because he happened to care a whole lot more about depression than schizophrenia.

Even in the earliest prescientific days of psychiatry, madness and melancholy were considered distinct conditions. Madness seemed to consist of disruptions of cognition, while depression seemed to consist of disruptions of emotion. There was certainly no medical or pharmaceutical reason to believe that a variation of a compound that dampened the hallucinations of psychotic patients could somehow increase the joy of depressed patients. Indeed, most psychiatrists believed that both psychosis and depression were the result of unresolved emotional conflicts. But Kuhn had quietly developed his own ideas about depression.

Kuhn rejected the standard psychoanalytic explanation that depression stemmed from repressed anger toward one's parents, so he rejected psychoanalysis as a treatment. Instead, he had become convinced that depression was the consequence of some kind of biological disturbance in the brain. Since nobody knew how chlorpromazine worked anyway, why not try a chlorpromazine knockoff on a depressed patient and see what happened?

So Kuhn gave G 22355 to three of his patients suffering from severe depression. He waited a few hours, then checked his patients. No improvement. He examined them again in the morning. Still nothing. Since chlorpromazine itself usually produced noticeable improvements within hours or even minutes of administration, it would have made sense for Kuhn to abandon his trials. Instead, for reasons known only to Kuhn, he kept on administering G 22355 to the three patients. Finally, six days after he had started the treatments, on the morning of January 18, 1956, one of these patients, a woman named Paula I., woke up and told the nurse that she was apparently cured of her depression.

Delighted, Kuhn contacted Geigy and announced that G 22355 "has an obvious effect on depression. The [condition] visibly improves. The patients feel less tired, the sensation of weight decreases, the inhibitions become less pronounced and the mood improves." In other words, Kuhn had just handed Geigy a silver platter containing what might very well be the world's first antidepressant. Did the Geigy executives uncork the champagne? Nope. They couldn't care less about depression. They wanted their own antipsychotic drug to compete with chlorpromazine. They ordered Kuhn to stop testing G 22355 and supplied the compound to another doctor, explicitly instructing him to test it out only on psychotic patients.

Kuhn tried to take his discovery to other scientists. In September 1957, Kuhn was invited to speak at the Second World Congress of Psychiatry and presented a paper about the effects of G 22355 on depressed patients. Barely a dozen people showed up. No one asked a single question. Frank Ayd, an American psychiatrist and devout Catholic who attended the talk, later reported, "Kuhn's words, like those of Jesus, were not appreciated by those in positions of authority. I don't know if anybody in that room appreciated we were hearing the announcement of a drug that would revolutionize the treatment of mood disorders."

It looked as if G 22355 was headed for the dustbin of history. But then an influential Geigy stockholder named Robert Boehringer happened to ask Kuhn if he could recommend anything for his wife. She was ill with depression. Kuhn immediately recommended G 22355. Boehringer's wife recovered. Watching the remarkable improvement in his wife, Boehringer lobbied Geigy to begin marketing the drug. In 1958, Geigy finally began marketing G 22355, naming it imipramine.

Imipramine became the prototype for dozens of antidepressant drugs that soon followed. Even today, every known antidepressant still shares the same basic mechanism as imipramine, influencing the neurotransmitter serotonin. Even Prozac is a tweaked version of imipramine. Though we still do not have a clear understanding of how antipsychotics or antidepressants bring about their improvements in mentally ill patients, we have a basic awareness of their physiological activity. Both chlorpromazine and imipramine are like shotgun blasts that spray everything in sight, rather than a sniper rifle that hits a single precise target. Chlorpromazine activates at least a dozen different types of neural receptors. Most of these have nothing to do with schizophrenia. It is hypothesized that the antipsychotic effects of chlorpromazine are produced by blocking two or three types of dopamine receptors. But if that was all the drug did, it would produce intolerable side effects, including severe involuntary movements known as dyskinesia. But chlorpromazine, and the many antipsychotic derivatives of chlorpromazine, also block serotonin receptors, a neural effect that fortuitously appears to mitigate the dyskinesia produced by the dopamine receptor blockade. This unusual interaction allows the drug to treat schizophrenia without producing intolerable side effects.

Imipramine also hits many different receptors in the brain, most of which have nothing to do with depression and several of which produce undesired side effects. But one of the targets of imipramine (and every known antidepressant) is the serotonin reuptake pump, which helps control the amount of the serotonin neurotransmitter in neural synapses. (Prozac and its analogs are known as "selective serotonin reuptake inhibitors," or SSRIs.) Why does increasing the brain's available serotonin reduce depression? We still don't know.

Why would two compounds that are very similar chemically each turn out to be an effective treatment for two very different mental disorders? There is a broad class of neurotransmitters that includes epinephrine, norepinephrine, and dopamine that are collectively known as biogenic amine substances, because they all share a particular chemical substructure known as ethylamine. This means that other molecules that contain an ethylamine substructure—even synthetic molecules that are not natural to the body—have a high probability of producing some kind of effects in the brain, or possibly multiple effects by activating different sites simultaneously. Scientists call a special chemical structure, like ethylamine, that can activate multiple targets in the body a "privileged structure."

Chlorpromazine and imipramine both contain an ethylamine substructure, which is why they have such broad and diverse effects on neural receptors in the brain. Purely by accident, Henri Laborit and Roland Kuhn had armed themselves with drugs that induced a wide range of changes in the brain, and merely got lucky that there were more positive changes than negative ones.

There's an old adage, "It's better to be lucky than smart." A drug hunter has the best chance of success when they are both lucky and smart—and Laborit and Kuhn were both.

Conclusion

The Future of the Drug Hunter

The Chevy Volt and the Lone Ranger

Is drug development more like engineering . . . or filmmaking?

"For success in drug hunting one needs 'four Gs': *Geld, Geduld, Geschick* . . . and *Glück*."

—Paul Ehrlich, 1900

In the fall of 2002, General Motors knew it was in trouble. It had predicted hybrid vehicles would never catch on with the public—after all, consumers loved GM's gas-guzzling SUVs, so there was little financial motivation to invest in the development of electric cars. But then, a bomb dropped. Toyota introduced the Prius, a gasoline-electric hybrid that quickly became a sales sensation and made Toyota the undisputed world leader in hybrid production. General Motors was suddenly staring into a future much different than what they expected or were equipped for.

Nevertheless, as with most industries that rely on engineering and science—including the computer industry, kitchen appliance industry, and telecommunications industry—the automobile industry usually offers a chance for a properly incentivized company to catch up to a market leader or, at the very least, carve out their own market share. All General Motors needed to do was design its own hybrid vehicle.

So GM assembled a team of their brightest scientists and

engineers and commanded them to build a vehicle that fulfilled two design objectives: first, the ability to drive across the entire country on gasoline; and second, the ability to commute to work without using any gasoline at all. Now let's pause here for one moment. Toyota had a head start of about ten years in terms of developing the Prius. General Motors, meanwhile, needed to figure out how to build their own version of an electricity-powered car from scratch. Nevertheless, even though nobody expected GM to design a vehicle as popular as the Prius, neither industry insiders nor ordinary consumers were especially skeptical that GM could build *some* kind of hybrid car. After all, the company had highly trained scientists and engineers who were well-versed in the technical knowledge required to achieve their goal. Their collective expertise covered battery technologies, electric motors, internal combustion engines, chassis engineering, and automobile design. They knew the manufacturing techniques for different parts and they knew the cost of materials.

So after eight years of effort, when GM's hybrid-design team rolled out the Chevy Volt—a vehicle that fulfilled both of its design objectives—it was certainly a triumph, but it was not a particularly astonishing achievement. After all, shouldn't GM, the world's largest automobile maker, know how to design a car?

In the end, the Volt did not sell very well, and it did not make a meaningful dent in the sales of Prius. But from an engineering standpoint, the sales are almost beside the point. The Volt did what it was supposed to do. In a relatively short period of time, General Motors had taken a vague design idea—let's create our own plug-in hybrid!—and actually crafted a product that turned the idea into reality. Now think how different this process is from engineering a Hollywood movie.

In 2007, Disney director and producer Jerry Bruckheimer was riding high on the success of three consecutive *Pirates of the*

Caribbean films, each one a global smash hit. They thought they had figured out how to design a blockbuster, so when Bruckheimer bought the rights to a new movie he began putting it together according to the same design principles—namely, a supernatural action comedy script penned by the same writing team behind the *Pirates* trilogy, a huge special-effects-laden budget, some romance, a happy ending, and Johnny Depp in a scenery-chewing starring role. Disney agreed that these were the right ingredients for success and funded the project. Yet, even though the producers diligently followed the ordained formula for commercial movie production, the final product failed to fulfill its basic design objectives: making an audience laugh, applaud, and feel genuine thrills. Instead, *The Lone Ranger* became one of the biggest flops of the past decade.

Unlike the Volt, Disney's movie simply did not work. That's because as much as Hollywood wishes there was some kind of cinematic blueprint for success, filmmaking is ultimately an artistic process, involving moments of inspired creativity and an abundance of trial and error. It is simply impossible to predict whether a particular script will become a successful movie.

This leads us to our concluding question: is the process of creating a new medicine more like designing the Volt or more like designing *The Lone Ranger*? Put another way, is drug discovery more like scientific engineering or artistic creativity? More than a century and a half after the establishment of a scientific pharmaceutical industry, the answer is clear. The development of new medicines—including antibiotics, beta blockers, psychoactives, statins, antifungals, and anti-inflammatories—is far more like trying to craft the next *Avengers* than developing a new car . . . or a new cell phone, vacuum cleaner, or satellite.

We instinctively presume that major medications—like insulin, Prozac, or the birth control pill—were developed through a rational process of scientific engineering similar to how the Volt was designed. Big Pharma executives identify the need for a particular drug, assemble a team of crack scientists, hand them a list of objectives, shower them with money, and wait for them to crank out the desired medicine. This is, in fact, a fair representation of the process that drug companies follow to develop copycat versions of existing drugs. For instance, just as General Motors hungrily eyed the impressive sales of the Prius, the drug company Lilly envied the sales of surprise blockbuster Viagra and assembled a drug development team to design its own erectile dysfunction drug. The result was Cialis, which carved out its own enviable share of the male arousal market. But Cialis was not some original creation, like the Volt. It was a knock-off; more like the way the Lincoln Navigator was a badge-engineered knock-off of the Ford Expedition. Cialis worked by acting upon the same physiological mechanisms as Viagra (namely, inhibiting the PDE5 enzyme). Lilly did not figure out how to treat erectile dysfunction in a way that avoided the existing side effects with Viagra (such as flushing, head-aches, indigestion, nasal congestion, and impaired vision). Its scientists merely duplicated Pfizer's drug, but found a molecular tweak that prevented Lilly from violating Pfizer's patent and added enough of a change in effect to allow Lilly some room for differential marketing (Cialis lasts longer than Viagra). Cialis was not an engineering breakthrough; it was Viagra 2.0—or, really, Viagra 1.1.

Game-changing drugs are almost never developed the way General Motors designed the Chevy Volt or Steve Jobs invented the iPhone or the way that most transformative consumer products are developed. Steve Jobs was able to tell his engineering team, "Go create a new kind of computer that is a flat tablet with a touch screen that runs Apple software" and expect that they would be able to build it. (Whether it would sell well is an entirely separate question; what is important is that he could be confident it *could* be built in a reasonable time frame and work exactly as he intended.) But Disney can never be confident when it tells its own team, "Go create a movie that makes people guffaw, cry, and cheer." Similarly, drug companies can never be sure that they will get a drug that works the way they hope it will.

The reason is as simple as it is profound: there still are no clear scientific laws, engineering principles, or mathematical formulae that can guide an aspiring drug hunter all the way from idea to product. Even though there have been a number of advances that make different components of the drug hunting process more efficient—such as receptor theory, rational design, recombinant-DNA engineering, pharmacokinetic testing (evaluating how a drug is processed by the body from ingestion to elimination), transgenic animal disease modeling (genetically engineering an animal's DNA to mimic some aspect of human disease in order to test the drug on the animal instead of a human), high-throughput screening (the ability to rapidly evaluate thousands of compounds), and combinatorial chemistry (the ability to generate thousands or even millions of different chemical compounds in a single process in order to use them for testing)—these are more akin to IMAX projectors, surround sound, and improved CGI rather than a blueprint for engineering a drug.

There's another similarity between filmmaking and drug hunting. Hollywood professionals take big risks. If your movie turns into a hit, you will be rich, famous, and potentially shape culture. If your movie flops, on the other hand, you may be broke, notorious, and depressed, potentially hurting your chances of securing support for your next cinematic effort. If you want to try to make it in Hollywood, you need to be brave, vehemently optimistic, and endowed with a short memory so you can forget all the flops you were part of. Of course, some might say you need to be crazy or foolish to work in Hollywood. Most drug hunters I've met are brave and optimistic, though a few do qualify as crazy and foolish. Not too many fall in between these extremes.

Scientists who search for new medicines must expose themselves to hazards both conspicuous and unknown. Valerius Cordus died from a disease he contracted while searching the wilderness for new botanical drugs. James Young Simpson inhaled a variety of volatile organic liquids in his search for an ether replacement, including many toxic ones. I myself took an experimental drug that made me ill in the hope that my self-experimentation would bring a useful medicine to patients more rapidly. More severely, in 2016, a painkiller drug trial in France killed one man and critically injured five others; though the drug scientists appear to have done everything right and weren't hurt themselves, they are facing litigation and will likely never work again—though they will be haunted by the man's death for the rest of their lives.

What's truly amazing, however, is how much success we have had in coming up with important drugs. We have cured dozens of major diseases and have effective treatments for everything from diaper rash to headaches to diarrhea to athlete's foot. Even though the drug discovery process is highly random and relies more on individual artistry than rational design, we live in a world where we can expect to find medications for most of our ailments. If drug hunters are more like filmmakers than automobile engineers, then how can we account for this counterintuitive success?

The thing about trial and error is that if you keep on trying and keep on being willing to make errors, eventually you will find something that works. And the more drug hunters we have striving to make the next *Star Wars*, the greater the chance that one of them will prove to be a J. J. Abrams of pharmacology.

Nevertheless, the unyielding difficulty of developing new drugs remains one of the biggest sources of the high costs of our medication. The R & D costs for the pharmaceutical industry are far higher than those of other technology-based industries such as automobiles, computers, and consumer electronics. One reason is that so many product development efforts by Big Pharma end up with *bupkis,* sometimes after a billion dollars was spent. Another reason is because of the high costs of complying with the strict and extensive FDA regulations designed to ensure the safety of our drugs. In addition, because of patent law and the lengthy drug development process, drugs have a relatively short window of market exclusivity (often ten years or less), so any potential profits must be accrued over a limited period of time. But despite the significant impact of FDA regulations and brief patent protections, if pharmaceutical companies could depend on the same clarity and reliability of engineering that auto makers and consumer electronics enjoy, then there's little doubt that the price of drugs would come down dramatically. Instead, Big Pharma must price their few successful drugs to cover the immense costs from their myriad unsuccessful drugs.

The soaring cost of developing new drugs creates financial disincentives that prevent pharma companies from focusing on drugs that produce cures. Why? Because any medication that can resolve a medical condition all at once is a medication that does not need to be purchased over and over again, drastically limiting its profit potential. For example, as we saw, the economics of antibiotics are quite unfavorable to Big Pharma, since patients need only to take a single course of the drug to get better, and doctors tend to hoard new antibiotics anyway. Vaccines are even worse, financially speaking, since (in principle) a person may need to take the drug only one time their entire lives. Moreover, there is a relatively low barrier to entry for competitors to make vaccines. Since vaccines tend to be public health drugs, they are often developed through government programs, which further reduces their commercial profitability. Antifungals—cures for fungus-based pathologies—suffer from the same profit limitations as antibiotics, with the added problem that there are far fewer people afflicted with fungus-based diseases than bacteria-based diseases. Antivirals, like Tamiflu, also tend to offer the same economic disincentives as other infectious disease cures, though HIV antiviral drugs have turned out to be a Big Pharma–enriching exception, since AIDS patients need to take a cocktail of anti-HIV medications every day for their entire lives.

This is not to say that incompetence, a prioritization of short-term gains over long-term goals, and naked greed (as opposed to economic disincentives) fail to play a role in keeping drug prices high or preventing valuable medicines from coming to the market. Human foibles are certainly present among pharma executives as they are everywhere else. But at its core, the pharma industry is predicated upon the same kind of profound and irremediable uncertainty as Hollywood. At the same time, there is a tiny handful of big movie studios who—against the odds—somehow seem to put out an unbroken stream of quality products that unceasingly delight audiences. At the moment, they may be the only ones. This handful has achieved rare consistency by granting their writers and directors unparalleled creative freedom with relatively little interference from producers and executives. Perhaps if Big Pharma were willing to grant their scientists the same kind of creative control over the drug hunting process, we might see a pharma company issue their own string of *Toy Story*, *Wall-E*, and *The Incredibles*.

Their own string of world-changing Vindications.

Appendix

Classes of Drugs

Neuropharmacological Drugs

Autonomic nervous system drugs

- Muscarinic
- Cholinesterase inhibitors
- Adrenergic drugs

Serotonin drugs

Dopamine drugs

Antipsychotics

Antidepressants

Anxiolytics

Hypnotics and sedatives

Opioids

General anesthetics

Anti-epileptics

Neurodegenerative disease drugs

Cardiovascular drugs

Renal drugs for hypertension

ACE-type drugs for hypertension
Beta blockers and other anti-hypertensives
Digitalis and anti-arrhythmia drugs
Anticoagulants
Anti-cholesterol drugs

Inflammation and the immune system

Antihistamines and related drugs
Aspirin-like drugs and related drugs
Immune system suppressants
Asthma drugs

Hormone drugs

Thyroid drugs
Estrogens and progestins
Androgens
Adrenal corticoid drugs
Insulin and other drugs for diabetes
Drugs acting on bone formation and degradation

GI drugs

Drugs for acid reflux and ulcer
Drugs acting on bowel motility

Anti-infective drugs

Malaria drugs
Protozoal infection drugs
Helminth infection drugs
Sulfa drugs
Penicillins
Streptomycin-like drugs

Quinolones and related drugs
Other antibacterial drugs
Drugs for tuberculosis and leprosy
Antivirals and drugs for AIDS

Cancer drugs

Cytotoxic drugs
"Oncogene" selective drugs

Reproductive System Drugs

Contraceptives
Gynecological and obstetric drugs
Erectile dysfunction drugs

Ocular drugs

Dermatology drugs

Notes

Introduction: Searching the Library of Babel

11: James Young Simpson: On October 16, 1846, at Massachusetts General Hospital, William T. G. Morton demonstrated that patients for the first time could be temporarily rendered unconscious prior to a surgical procedure. The drug he used was ether. Today when a pharmaceutical company gets FDA approval for a new kind of drug, competing companies promptly start their own research programs to find a similar drug. These are often called "me-too" drugs. Chloroform might be the first "me-too" drug of the industrial era.

As a more modern example, shortly after Squibb got a new kind of drug named captopril approved for the treatment of hypertension, Merck set to work developing its own me-too anti-hypertensive drug, which became enalopril. Similarly, when Lilly received FDA approval for Prozac in 1987, Pfizer quickly followed up with the me-too antidepressant Zoloft, while GlaxoSmithKline obtained approval for the me-too Paxil.

Chapter 1: So Easy a Caveman Can Do It: The Unlikely Origins of Drug Hunting

19: If we relegate alcohol to the status of a beverage: There are good reasons for classifying alcohol as a foodstuff and not a drug. The discovery of

Stone Age beer jugs demonstrated that intentionally fermented beverages existed at least as early as 10,000 BC, and some historians even suggest that beer may have preceded bread as a food staple. Alcoholic potables were very important in ancient Egypt, and beer—commonly brewed at home—was considered a necessity of life. But alcohol was also used for medicine, as an offering to the gods, and was an important component of funerals; alcoholic drinks were frequently stored in the tombs of the deceased for their use in the afterlife. The Egyptian god Osiris was even believed to have invented beer, signifying the beverage's divine nature.

Nevertheless, for most of human history alcohol has been regarded as a cure-all medication. Spirits, whose alcohol content was increased through distillation, were commonly used for medical purposes, and many of the names for these drinks reflect old beliefs regarding their curative properties. "Whiskey" comes from the Gaelic word *usquebaugh* meaning "water of life," which is also the translation for *eau de vie,* what the French call non-barrel-aged distilled spirits. Legend has it that that when sick patients were dosed with these potent potables they would thrash about and become more animated, thus leading to the conclusion that the drink was putting life back into them. (This particular experiment can be repeated today in the privacy of your home.)

We now know that ethyl alcohol (the alcohol in fermented beverages) acts by stimulating GABA A (γ-aminobutyric acid) receptors. These receptors are the major neuro-inhibitory receptors in the brain, and stimulating these receptors causes a decrease in neurological activity, causing sedation. Benzodiazepines, commonly referred to as tranquilizers (which include Librium and Valium), target the same class of receptors. One common use of benzodiazepines is for the treatment of insomnia. I remember my grandmother on occasion treating her own insomnia by taking a little bit of schnapps before bedtime. Another common use for benzodiazepines is to treat the symptoms of anxiety.

This, in the end, points to the severe limitations of alcohol as a drug: the lack of separation between desirable therapeutic activity and undesirable side effects. Benzodiazepines are far more effective at treating anxiety because their effects are more targeted.

19: Opium is the active ingredient of the poppy: In this chapter we use opium as an example of a medicine that was found by ancient botanical drug hunts and still meets contemporary standards for a therapeutic drug. There are, however, several other examples. Ergot, for instance. Ergotamine and related compounds are produced by the ergot fungi in the genus *Claviceps*. These are pathogens that infect cereal plants, most commonly rye. When the fungus grows on the plant it produces ergotamine and a suite of other toxic compounds. In ancient times, these ergot-related compounds were ingested by humans when they ate rye produced from plants infected by the fungus. Today, we would call these compounds "dirty drugs" because they act simultaneously on multiple targets in the body. As a result, eating ergot produces symptoms that are quite varied and complex.

One class of symptoms is convulsions, including seizures, nausea, and vomiting. The second class of symptoms from ergot poisoning is hallucinations. Chemically speaking, ergotamine is very closely related to LSD. Finally, ergot poisoning can also elicit gangrene. Ergotamine is a potent vasoconstrictor, meaning that it narrows blood vessels. This reduces the body's blood supply, which poses particular difficulties for the peripheral regions of the body, such as hands, fingers, feet, and toes. The extremities initially feel a tingling sensation similar to the experience of "pins and needles" or a body part "falling asleep" when it is held in an awkward position for a period of time. Of course when that happens, you simply move your body and shake out the limb, restoring blood flow, making the numbness go away. That will not work if you are poisoned by ergot. Instead, the skin on the parts of your body feeling "pins and needles" will begin to peel off. Eventually, your appendages will swell, blacken, and die—"falling asleep" permanently.

Epidemics of ergot poisoning occurred periodically throughout history. The unexplained appearance and just-as-sudden disappearance of the pestilence together with the hallucinations and the blackening and dying of toes and fingers made it easy to imagine that the condition was caused by evil possession or the wrath of God. One early account of ergot poisoning appeared in the *Annales Xantenses* in 857: "a great plague of swollen blisters consumed the people by a loathsome rot, so that their limbs were loosened and fell off before death." In the Middle Ages, ergot poisoning was called Saint Anthony's Fire, after monks of the Order of Saint Anthony discovered a cure for the condition. How did these benighted medieval monks manage to produce a cure? Through prayer and penance—literally. When victims were afflicted with ergot poisoning, they would go to monasteries to perform prayers and penance and ask God for mercy. Medieval monasteries did not grow rye, however—they grew wheat and barley instead. So as long as a victim stayed at the monastery, they stopped eating the contaminated rye, causing the symptoms to recede. Of course, the recovered penitent would return home and resume eating bad rye and the symptoms would reappear. The monks explained the resurgence of Saint Anthony's Fire by suggesting that the wayward Christian had returned to their lax and immoral ways, incurring the wrath of God once again. Naturally, a return to the piety of the monastery would set things right, morally and physically.

Another ancient drug that is still in use today is digitalis, which remains the drug of choice for many cardiac patients. Plant extracts containing digitalis compounds were used in primitive societies as arrow poisons. One of the first digitalis drugs is mentioned in the Ebers Papyrus, a document containing Egyptian herbal knowledge written around 1550 BC, which means the Egyptians were using the plant extract more than 3,500 years ago. Digitalis is also mentioned in writings by Welsh physicians from 1250 AD. The plant from which digitalis is extracted, foxglove, was described botanically by Fuchsius in 1542 who named it *Digitalis purpea* based on the appearance of the flower, which is purple in color and is said to resemble a human finger.

The therapeutic value of digitalis was fully described by the physician William Withering in a 1785 book entitled *An Account of the Foxglove and Some of Its Medical Uses: with Practical Remarks on Dropsy and Other Diseases.* Withering described how he first came to use digitalis some ten years prior to the publication of his treatise:

> In the year 1775, my opinion was asked concerning a family in receipt of the cure of the dropsy. I was told that it had long been kept a secret by an old woman in Shropshire, who had sometimes made cures after the more regular practitioners had failed. I was informed also, that the effects produced were violent vomiting and purging: for the diuretic effects seem to have been overlooked. This medicine was composed of twenty or more different herbs: but it was not very difficult for one conversant in these subjects, to perceive, that the active herb could be none other than the Foxglove.

Dropsy is an archaic term for the swelling of soft tissues due to the accumulation of excess water, today called edema. It is commonly seen in heart failure. Withering was an expert botanist as well as a physician, which enabled him to recognize that foxglove was likely the active ingredient in the complex mixture promoted by the Shropshire woman. Even so, Withering did not appreciate that the primary benefits of foxglove were due to its actions on the heart, though he did recognize that foxglove produced cardiac effects, writing:

> It has a power over the motion of the heart to a degree unobserved in any other medicine, and this power may be converted to salutary ends.

Despite Withering's clear description of both the benefits and harmful side effects of foxglove, the drug was used indiscriminately throughout the nineteenth century for a wide variety of diseases, often in doses that were toxic. During the early twentieth century, the drug came to be used specifically for atrial fibrillation (an irregular and rapid heart rate) and in the middle of the twentieth century it was finally appreciated that the main therapeutic value of digitalis is as a treatment for congestive heart failure. A damaged heart muscle will work more efficiently in the presence of digitalis, restoring the health of a patient following a heart attack. In order to achieve this effect, digitalis must be dosed with great precision, as even a slight overdose will make the patient's condition worse rather than better.

One final drug that was used during ancient times and that still has modern value is colchicine, a treatment for gout. Gout is a painful inflammatory disease that results from the deposit of uric acid crystals in the joints, most commonly in the joints of the large toe. Ancient Egyptians first described gout in 2,600 BC as a kind of arthritis of the big toe. It was often referred to as a "rich man's disease," because there is a strong association between gout and the consumption of alcohol, sugary drinks, meat, and seafood, foods that were once limited to the wealthy classes. An English physician named Thomas Sydenham penned an early description of the disease in 1683:

> Gouty patients are, generally, either old men, or men who have so worn themselves out in youth as to have brought on a premature old age—of such dissolute habits none being more common than the premature and

> excessive indulgence in venery, and the like exhausting passions. The victim goes to bed and sleeps in good health. About two o'clock in the morning he is awakened by a severe pain in the great toe; more rarely in the heel, ankle or instep. The pain is like that of a dislocation, and yet parts feel as if cold water were poured over them. Then follows chills and shivers, and a little fever. . . . The night is passed in torture, sleeplessness, turning the part affected, and perpetual change of posture; the tossing about of body being as incessant as the pain of the tortured joint, and being worse as the fit comes on.

Many modern treatments for gout focus on eliminating the uric acid crystals that cause the painful symptoms. Since gout is an inflammatory disease, anti-inflammatory drugs are often effective in relieving its symptoms, such as ibuprofen. Colchicine is extracted from the seeds and tuber of *Colchicum automnale*, also known as the autumn crocus or meadow saffron. *Colchicum* plant material was recommended for the treatment of rheumatism and swelling in the Ebers Papyrus, and for the treatment of gout by Greek physician Alexander of Tralles in the middle of the fifth century AD. Benjamin Franklin suffered from gout and is said to have been responsible for introducing colchicum to the American colonies. Interestingly, since cochicine has been used for so very long, no one ever bothered to seek Food and Drug Administration approval for it as a stand-alone gout treatment until 2009, about 3,500 years after its clinical introduction.

20: Dover's Powder: In 1709, Dover's expedition landed at a deserted island off the coast of Chile that was part of the Juan Fernandez archipelago. There they found Alexander Selkirk, a man from Largo, Scotland, who had escaped to sea to in 1703 in order to avoid a command to appear in court for "undecent carriage" (indecent behavior) in church. Selkirk had sailed from England on the privateer *Cinque Ports* but abandoned the ship for the remote island in 1704 following a dispute over the vessel's seaworthiness. Indeed, the *Cinque Ports* later foundered, losing most of its hands. Selkirk was wise to get off the ship, but even so, he was stranded for more than four years until his rescue by Dover's expedition. Selkirk became a minor celebrity in England with his story being recounted in a much-read article that appeared in the magazine, "The Englishman," eventually inspiring Daniel Defoe to write *Robinson Crusoe.*

Chapter 2: Countess Chinchón's Cure: The Library of Botanical Medicine

37: A year after Talbor's death: Today, most problems of drug misrepresentation occur around issues pertaining to off-label promotion—marketing drugs for uses not approved of by the FDA—or from pharma companies providing kickbacks to physicians as compensation for prescribing the company's drugs. Recent lawsuits that involved off-label promotion have included a $2.2 billion legal settlement by Johnson and Johnson in 2013, a $3 billion settlement by Glaxo Smithkline in 2012 (of which $1 billion was for criminal charges), and a $2.3 billion Pfizer settlement in 2009.

Chapter 3: Standard Oil and Standard Ether: The Library of Industrial Medicine

44: Dr. John Warren, one of the founders of Harvard Medical School: John Warren was born in Roxbury, Massachusetts, near Boston on July 27, 1753. He was the youngest of four brothers. His father, also Joseph Warren, was an apple farmer and Calvinist who strongly instilled within his sons the value of higher education and love of country. John, Jr. did well in grammar school and attended Harvard, entering in 1767 at the ripe age of fourteen years. At Harvard he learned Latin, became a competent classical scholar, and developed a strong interest in the study of anatomy, joining the Anatomical Club, where he dissected lower animals and studied the human skeleton. Studying human bones was not easy, as there was no ready supply of corpses. In order to pursue his studies, Warren along with his fellow students kept a close watch on the disposal of dead criminals and vagrants.

After graduation, Warren started a medical practice in Salem, Massachusetts, in 1773. Warren's practice was greatly affected by the Revolutionary War, and his brother Joseph was killed during the battle of Bunker Hill. In 1780, John Warren was among the first to propose the establishment of a medical school as part of Harvard University. By 1782, Harvard had established three professorships in medicine, and Warren was appointed chairman of the anatomy department at the new medical school.

One of his students, James Jackson, wrote that one of the "most peculiar charms" of Warren's teaching was "derived from the animation of delivery, from the interest he displayed in the subject of his discourse and from his solicitude that every auditor be satisfied both by his demonstrations and by his explanations." Warren developed a reputation as an excellent surgeon, highly regarded for pioneering new surgical procedures. By 1815, Harvard Medical School had fifty students, and John Warren's oldest son, John Collins Warren, served as adjunct professor in anatomy and surgery. Thirty years after that, this adjunct professor performed the very first operation conducted under anesthesia.

Chapter 4: Indigo, Crimson, and Violet: The Library of Synthetic Medicine

67: Salicylates had been used for thousands of years: Most plants produce salicylates, which are hormones that enable different parts of the plant to communicate with each other. The willow tree happens to be a particularly rich source of these compounds but is otherwise unremarkable in salicylate physiology. An example of the operation of salicylates can be seen when a plant is afflicted with a disease known as systemic acquired resistance (SAR), a condition that occurs when one part of the plant is infected by a virus or fungus. If you try to infect a different part of the plant with the same pathogen a day or two after the initial infection, the plant will now be resistant to the pathogen. Why? The infected part of the plant releases salicylates into its vascular system, where it circulates to other parts of the plant and triggers the production of toxins that are kind of like plant antibodies (known as resistance factors) at these new sites, which help control the spread of the infection.

Plant resistance factors can be very toxic and can even cause an entire part of a plant to die off; this is not a defense strategy that can be used by animals, as losing an arm or leg is likely a worse state of affairs than dealing with a disease. But since a plant can survive the loss of one of its branches or roots, it is an excellent survival strategy for flora. This is why you often see a tree or bush with a dead limb. Animals on the other hand, including humans, possess an analogous defense system called innate immunity. One of the reasons we feel so crappy when we are sick with a cold or the flu is that our innate immunity system has kicked in and is producing chemicals that are toxic to the pathogen—but also toxic to ourselves.

70: Their antibiotic eventually received FDA approval and is currently generating over $1 billion in annual sales: Another example of a poor decision to cut bait instead of continuing to fish can be found in the hunt for anticholesterol drugs. In 1975, driven by new discoveries in biochemistry, Merck stepped up its efforts to study the synthesis of cholesterol in the body. It was known that HMG-CoA reductase was the first enzyme in the body's pathway for synthesizing cholesterol, and so Merck scientists began to search for compounds that inhibited HMG-CoA reductase. They speculated that such inhibitors might prove to be effective anticholesterol drugs. After just one week of testing a few hundred random samples, the researchers detected a very strong HMG-CoA reductase inhibitor candidate. This was an exceptionally speedy triumph, since it commonly takes several thousands of samples to find a good candidate. In 1979, the Merck scientist Carl Hoffman purified the inhibiting compound, giving rise to one of the earliest successful statin drugs, lovastatin. The drug was approved by the FDA in 1987 as a standard treatment for hypercholesterolemia.

At this point, Merck had a head start on other drug companies in the search for other viable anticholesterol drugs. Since lovastatin came from a soil microorganism, Merck reasoned that the library of dirt was the best place to find better statin drugs, even though they had already made progress in searching the library of synthetic molecules for HMG-CoA inhibitors. Merck decided to cut bait and curtail their efforts to develop a better anticholesterol medication through synthetic chemistry and instead focus solely on soil-based compounds.

Sensing opportunity, Merck's competitor Warner Lambert picked up on the chemistry line of HMG-CoA inhibitor research that Merck had abandoned and discovered an even better inhibitor—Lipitor. Lipitor quickly and dramatically outpaced the sales of lovastatin (and another one of Merck's soil-based statin drugs, simvistatin).

Chapter 6: Medicine That Kills: The Tragic Birth of Drug Regulation

98: Bayer dubbed their new drug Prontosil: Even though the discovery of Prontosil was based on a totally false hypothesis (the notion that dyes could be effective antimicrobial drugs), there was no arguing with its success. Gerhard Domagk, the Bayer research director who led the Prontosil research team, was awarded the 1939 Nobel Prize in Medicine. Unfortunately for Domagk, he did not get to enjoy his prize for long. The earlier award of the 1935 Nobel Peace Prize to Carl von Ossietzky, a German who was very critical of the Nazis, had so angered the German government that the Nazis made it illegal for any German to accept a Nobel Prize. Domagk was forced by the Nazi regime to refuse the prize, and he was eventually arrested by the Gestapo and jailed for a week.

105: the Food and Drug Administration oversees the development of drugs from the very beginning: Regulatory authorities have avoided providing a defined checklist of required experiments for clinical testing and instead issue general guidance notes for the required studies. Despite the reluctance of the FDA to share explicit guidelines, in practice the required studies are easily described via a checklist:

- Testing for Acute Toxicity: The drug is administered to a laboratory animal, usually a rodent, in increasing doses and the animal is observed for toxic effects following each dose. The dose range is large; from very low doses to the highest well-tolerated level (called the "no toxic effect level") and up to higher doses that produce obvious toxicity. At the end of each experiment, the animals are sacrificed and autopsied to search for any effects the drug may have had on their internal organs.
- Testing for QT Interval Prolongation: It is well known that some pharmaceutical targets can be very difficult to inhibit, some are relatively easy to inhibit, and some are so susceptible to the influence of drugs that they are sometimes unintentionally inhibited by drugs designed to act elsewhere. One of these highly susceptible targets is the cardiac hERG channel, an ion channel involved in regulating the rhythmic action of the heart. Inhibition of the hERG channel causes prolongation of the QT interval in the heart rhythm, which can lead to a potentially fatal heart arrhythmia called torsades de pointes. Drugs from many different therapeutic classes, including

tricyclic antidepressants, antipsychotics, antihistamines, and antimalarials, all inhibit the hERG channel. hERG channel inhibition must therefore be measured prior to the initiation of clinical trials.

- Testing for Genotoxicity: Cancer is caused by gene mutations that can either be inherited or produced during the course of life by exposure to certain viruses, radiation, or mutagenic chemicals. It is therefore crucial to avoid producing drugs that have any mutagenic activity, as such drugs could be carcinogenic. Bruce Ames, one of the scientists responsible for our current understanding of the mutagenic nature of carcinogenesis, developed a straightforward bacterial-based test, named the Ames test in his honor, to detect mutagenic activity and thus carcinogenic potential in any chemical compound. The FDA requires Ames testing as part of NDA testing, as well as related tests for chromosomal abnormalities and chromosomal damage in rodents.
- Testing for Chronic Toxicology: The acute toxicity study looks for immediate damage produced by a drug. However, there is also the concern that toxicity may occur when the drug is dosed repeatedly over time, even at low doses. Chronic toxicology studies address this concern. Three (or more) doses of a particular drug are administered over an extended period of time: a dose known to be toxic (from the acute toxicity study), a therapeutic dose, and an intermediary dose. This test is carried out on two species: a rodent (usually rats) and a non-rodent (usually dogs, though monkeys and pigs are used in certain circumstances). The duration of the chronic toxicity trial must match the intended clinical use. A two-week trial is adequate for a compound like an antibiotic that will be administered only for a few days. Studies of six months or more in duration are required for drugs that will be chronically administered, such as high blood pressure medications. Such studies can obviously be extraordinarily expensive because they must be conducted over a long period of time and require a large number of animals (100 rats and 20 dogs would be typical); in addition, the test requires large amounts of the actual test drug, which must be prepared in a costly fashion in order to meet the exacting FDA standards.

The reason it is so expensive to manufacture drugs for FDA testing is that they must be prepared under "good manufacturing practices" (GMP) guidelines. The synthetic process must be clearly defined and described in detail and followed precisely from batch to batch. Analytical procedures must be developed and validated to ensure quality control of the manufactured drug. The purity of the drug must be determined, and any impurities present must be defined, characterized, and rendered consistent from batch to batch. In addition, the drugs are not just administered to test subjects in a straightforward manner, but are dosed in a complex formulation that optimizes the delivery of the drug. This formulation must be optimized and remain fixed for all future studies. Finally, these studies must be carried out in specialized GMP laboratories operated under close regulatory oversight.

Many, many things can go wrong during the course of such toxicity studies. I remember one set of FDA tests in which everything was going well until we saw gastrointestinal bleeding in rats during the chronic toxicology study. We were astounded. There was absolutely nothing about the biological activity of the compound that would predict gastrointestinal bleeding, and we had not seen anything like that in prior experiments. After a long and expensive investigation we learned that the compound was crystallizing into long sharp needles when exposed to the acidic conditions of the stomach. Over time, the sharp needles accumulated and began to rip up the gastrointestinal tract lining. It was a physical and not a biochemical effect, but even so it stopped our FDA tests dead and sent us reeling back to the drawing board.

106: groups like ACT UP petitioned the FDA to loosen its criteria for the clinical testing: Probably no incident tilted public opinion more toward the side of caution than the thalidomide disaster. This drug was first developed in 1953 by the Swiss pharmaceutical firm Ciba. The company soon discontinued its research on thalidomide because they were unable to demonstrate any clear pharmacological effects. Despite this, another firm, Chemie Grünenthal in Stolberg, Germany, took over the development of the drug, and thalidomide was introduced to the public on October 1, 1957. It was initially marketed as an anticonvulsant but quickly proved ineffective for this purpose. Thalidomide chemically resembles barbiturates, and the company scientists probably thought that it could work similarly and thus be effective against epilepsy, but they obviously never checked to see whether it worked the same way as barbiturates. It doesn't—not at all.

Although the drug failed as an antiepileptic, it was noted that thalidomide produced a deep sleep without a hangover. In addition, large doses were not fatal, so unlike other sedatives, thalidomide posed no suicide risk. Thalidomide soon became the most popular sleeping pill in West Germany, where it was widely used in hospitals and mental institutions. It was marketed for the treatment of a variety of conditions including influenza, depression, premature ejaculation, tuberculosis, premenstrual symptoms, menopause, stress headaches, alcoholism, anxiety, and emotional instability. By the end of the 1950s, thalidomide was marketed by fourteen pharmaceutical companies in forty-six countries.

Thalidomide was also found to be an effective antiemetic (anti-nausea drug) and thus was prescribed to thousands of pregnant women to relieve the symptoms of morning sickness. At the time, it was believed that most drugs could not pass from the mother to the child across the placental barrier and thus there was little concern that the drug could harm the developing fetus. However, in the late 1950s and early 1960s there was a surge in the number of children with birth deformities, especially phocomelia (flipper-like arms or legs). In total, there were more than 10,000 such reports from all forty-six countries where thalidomide was sold. On opposite sides of the world the Australian obstetrician William McBride and the German pediatrician Widukind Lenz independently hypothesized a link between thalidomide and these birth defects, a link that was convincingly demonstrated by Lenz in 1961.

The impact in the United States was minimal because of the actions of Frances Oldham Kelsey, an FDA reviewer who refused to approve thalidomide. There had been reports of peripheral neuropathy associated with thalidomide use, and she insisted that additional testing was needed prior to FDA approval. Kelsey also noted that the manufacturer had provided only minimal animal safety data and that long-term risk assessment and pregnancy risks assessments had not been performed. Consequently, thalidomide never went on sale in the USA in the 1950s or 1960s.

However, it should be pointed out that no drug is ever all good or all bad, but depends heavily on the dose, individual, and context. For years after it was first prescribed, nobody knew how thalidomide actually worked. University studies eventually showed that thalidomide is a useful treatment for erythema nodosum leprosum (ENL), a painful complication of Hanson's disease, more commonly known as leprosy. In 1991, Gilla Kaplan at Rockefeller University showed that thalidomide worked in leprosy by inhibiting tumor necrosis factor alpha (TNF alpha). TNF alpha is a cytokine, a hormone that regulates immune cells, induces inflammation, and inhibits tumorigenesis and viral replication. Further work by Robert D'Amato at Harvard Medical School showed that thalidomide was a potent inhibitor of new blood vessel growth and this finding suggested that thalidomide could be used as a cancer treatment. In 1997, Bart Barlogie reported that thalidomide was an effective treatment for multiple myeloma, and soon thereafter the FDA approved thalidomide for the treatment of this cancer as well as for the treatment of leprosy. However, before receiving thalidomide, patients must go through a special process to prevent the drug from producing birth defects. Although the FDA feels that appropriate precautions are in place, the World Health Organization (WHO) has stated that:

> The WHO does not recommend the use of thalidomide in leprosy as experience has shown that it is virtually impossible to develop and implement a fool-proof surveillance mechanism to combat misuse of the drug.

Chapter 8: Beyond Salvarsan: The Library of Dirty Medicine

133: the world's first expanded-spectrum antibiotic: Benzylpenicillin has a fairly good antibacterial spectrum but is not considered to be broad-spectrum. Penicillin was the world's first true antibiotic, full stop.

134: Though penicillin was a true miracle drug, some bacteria-borne diseases remained impervious to it: Penicillin was not a perfect drug. After its success, many antibiotic research programs were initiated to make improved versions of penicillin. Some of the goals of these programs included finding compounds with a broader spectrum of antibiotic activity, finding compounds that could be given orally instead of by injection (benzylpenicillin cannot be given orally), finding compounds that could also fight bacteria in the central nervous system (penicillin compounds generally cannot pass the blood-brain barrier into the central nervous system and thus cannot be used to treat brain infections such as bacterial encephalitis), and most importantly, finding compounds that could reduce or overcome bacterial resistance. These research programs often focused on discovering naturally occurring penicillin-like chemical scaffolds for chemical synthesis. These penicillin-like chemicals include a specific molecular feature called the beta-lactam ring. The beta-lactam ring is usually drawn as a square in the middle of the penicillin molecular structure and serves as the compound's "warhead," producing the drug's toxic attack on bacteria. Some examples of compounds with a beta-lactam ring include cephalosporin, monobactams, and carbapenam.

134: Perhaps the most dreadful of these diseases was tuberculosis: Though tuberculosis is no longer much of a concern in the developed world, it is estimated that about one in three humans alive today is infected by *Mycobacterium tuberculosis*, with new infections occurring at the rate of about one per second. Though most tuberculosis infections are asymptomatic and harmless, there are currently about fourteen million chronic active cases worldwide with about two million deaths annually.

134: known as the "White Death": Tuberculosis also came to be known as the "great white plague" and was called "white" because of the extreme anemic pallor of those afflicted. The American physician and man of letters Oliver Wendell Holmes actually coined the term "great white plague" in 1861 when he was comparing the consumption epidemic to other horrific diseases of the era. Though tuberculosis patients famously took on a deathly pale hue, some historians suggest that the term "white" may have referred to the disease's cultural association with youth, innocence, and perhaps even saintliness, since afflicted patients took on a quasi-angelic appearance—opalescent, ethereal, almost delicate. Some writers with a more literary (or misogynistic?) bent suggested that the wan faces of its female victims rendered them particularly attractive, with at least one male observer declaring that the disease endowed women with a "terrible beauty."

137: Waksman would eventually receive a Nobel Prize: Waksman received the Nobel Prize in Medicine in 1952 for his work developing streptomycin as a cure for tuberculosis, but his collaborator Albert Schatz was not named on the Nobel Citation. This exclusion was strongly contested by Schatz, eventually leading to litigation. Waksman settled out of court, providing financial remuneration to Schatz and stating that Schatz was entitled to "legal and scientific credit as co-discoverer of streptomycin."

142: fifteen of the eighteen largest pharmaceutical companies have abandoned the antibiotic market entirely: About 99 percent of all microorganisms that live in the soil will die if you try to grow them on a Petri plate. This has always been a limiting factor when searching for new drugs in the library of dirt. But in the early 2000s, two professors at Northeastern University, Kim Lewis and Slava Epstein, figured out how to culture microorganisms that previously had been thought to survive only in the soil. After this technological breakthrough, it was suddenly possible to study and develop these so-called "unculturable bugs" for the first time.

Lewis and Epstein started a new company in Cambridge, Massachusetts, called NovoBiotic Pharmaceuticals to find new antibiotics using their new method. But even though they were able to grow soil microorganisms that had never before been culturable on a Petri dish, their basic approach was the same as previous excursions into the library of dirt: they randomly grew any microorganisms they found in the soil and screened them to see if they produced chemicals that would kill pathogenic bacteria.

In early 2015, NovoBiotic Pharmacticals reported finding an important new antibiotic known as teixobactin. Teixobactin appears to be active against many highly drug-resistant pathogens while remaining safe in animals.

Chapter 9: The Pig Elixir: The Library of Genetic Medicine

151: Indian physicians observed that ants were attracted to the urine: Since diabetes mellitus roughly translates to "excessive sweet urine," it should not be difficult to imagine what test was used to confirm the diagnosis prior to the twentieth century. Tasting urine sounds disgusting and potentially dangerous, but prior to the development of modern biochemical instruments, dipping your tongue into a patient's urine was both commonplace and useful. Early scientists did many things that would be considered foolhardy or risky today. For example, the lab notebooks of Louis Pasteur, the late nineteenth-century microbiologist, reveal that he frequently tasted the results of his biochemical experiments. Marie Curie died at the age of sixty-six from aplastic anemia, almost certainly caused by her exposure to the radioactive chemicals she studied all her life. Even today, Curie's notebooks are still considered too dangerous to handle because of their high levels of radioactivity.

These historic artifacts are kept in lead-lined boxes, and historians who wish to consult them must wear protective clothing. When I was first trained in chemistry about forty years ago, I was taught to sniff the chemicals that I worked with in order to determine whether my intended chemical reactions had proceeded properly. Fortunately, such hands-on—and nose-on—instruction is absent from the twenty-first-century chemistry classroom.

153: Whenever researchers ground up a pancreas in the hope of extracting insulin: Two physicians, Frederic Allen and Elliott Joslin, were among the most recognized experts in the treatment of diabetes in the early twentieth century. At the time, the major goal for diabetes treatment was to reduce the level of glucose in the blood. But since there was no access to insulin, the best that physicians could do was attempt to reduce the glucose levels in a patient's diet. Unfortunately, animal experiments eventually demonstrated that diabetes is not merely a problem with glucose metabolism but also a problem with metabolizing protein and fat. If one simply removes carbohydrate from the diet, the body will burn fat and protein instead, producing chemicals called acidic ketone bodies, which acidify the blood. The blood's pH (the measure of acidity in a solution) must be maintained in a very narrow range near neutrality, between pH 7.35 and pH 7.45. Acidosis, or the lowering of blood pH, results in respiratory distress, heart arrhythmia, muscle weakness, gastrointestinal distress, coma and, if untreated, death.

Thus, in the absence of insulin, Allen and Joslin's only available treatment for diabetes was to completely starve the patient, removing all carbohydrate, protein, and fat from the diet. Of course, you cannot live without eating *something*, so Joslin and Allen developed a diet that provided about 20 percent of the calories ordinarily needed for survival and in a form that was particularly low in carbohydrates and sugar. Such a diet minimized collateral damage to the patient's cells, but still produced severe emaciation. Joslin described one of the dieting patients in his Boston clinic as "just about the weight of her bones and a human soul." The radical diet was not a cure, but it could extend life somewhat. One might naturally ask, however, what was the use of extending life if the quality of life was so miserable that you were both ravenous and without energy for any of the normal activities of life. The only rationale for sticking to such a desolate diet was to try to survive until a real cure could be found.

157: Collip applied state-of-the-art biochemistry techniques to refine the insulin: Even though all proteins share several physical properties, they often differ in their solubility in alcohol. Collip therefore explored a technique known as alcohol precipitation fractionation as a means to purify insulin. This meant slowly adding alcohol to an impure insulin compound until the point at which the insulin just barely remained soluble. All the other proteins in the dirty compound that were less soluble than insulin would then precipitate, forming tiny particles in the liquid that could easily be removed.

162: Paul Berg, a Stanford University professor who studied viruses, performed one of the most important experiments of the twentieth century: Berg teamed up with two other professors working in the San Francisco Bay area to optimize the new recombinant DNA techniques: Herb Boyer at the University of California, San Francisco—an expert on the enzymes that cut and pasted DNA—and Stanley Cohen, also at Stanford, who was an expert on plasmids, tiny circles of DNA that are natural carriers of genes between organisms.

Chapter 10: From Blue Death to Beta Blockers

184: You are likely familiar with adrenaline's role in the fight-or-flight response: The May 13, 2010, issue of the *New England Journal of Medicine* reported the case of a fifty-four-year-old woman admitted to Massachusetts General Hospital following multiple episodes of dizziness, sweating, and palpitations that caused her to fall. Upon examination she was found to be suffering from high blood pressure. However, her blood pressure varied greatly depending on her posture: it went up when she was sitting or lying down, but would drop significantly when she was standing or walking, causing her to faint when the blood pressure fell to an extremely low level.

Eventually, the source of the patient's high blood pressure was determined to be a rare type of tumor of the adrenal gland called a pheochromocytoma. This tumor secretes large amounts of adrenaline. Anyone who has been in a car accident or a near-accident knows what an "adrenaline rush" feels like. Your heart races, everything seems to slow down, and you feel hyperaware of your surroundings. In addition, your blood pressure goes up. All of this occurs because your adrenal gland rapidly releases a large quantity of adrenaline in moments of perceived danger.

In most patients with a pheochromocytoma, large amounts of adrenaline are produced all the time, raising their blood pressure all the time. But in some cases, like this woman, a patient's body can adapt to the sustained onslaught of adrenaline, resulting in more variable levels of blood pressure. When the woman was lying down with her heart at the same level as her head, her blood pressure remained high and she was able to circulate adequate amounts of blood to her brain. Normally, when you sit up the circulatory system rapidly compensates for the fact that your head is now at a higher level than your heart by increasing blood pressure in order to maintain steady blood circulation to the brain. But in this woman suffering from a pheochromocytoma, her body overcompensated for the high levels of adrenaline and could not maintain the pressure necessary to circulate adequate blood to the brain, leading to her fainting spells.

The patient underwent surgery to remove the tumor. Afterward, her level of adrenaline declined dramatically, as did her dizziness, and she was able to return to work.

184: Armed with this promising idea, Black approached the British company ICI Pharmaceuticals: Black used a similar strategy to develop a treatment for stomach ulcers, but ICI was not interested in ulcer medicines, so eventually Black resigned and in 1964 joined Smith, Kline and French in order to pursue his stomach ulcer research at their labs. His work there culminated in the discovery of cimetidine (Tagamet), which soon after its launch in 1975 turned into another blockbuster drug. It became the first drug in history to reach $1 billion dollars in annual sales.

186: an enzyme in the body known as ACE: The kidney produces and secretes a protein called renin in response to low blood pressure, which initiates a chain of events leading to increased blood pressure. In the bloodstream, renin cleaves a peptide (a very small type of protein) that is produced in the liver called angiotensinogen to produce an even smaller peptide called angiotensin I. Angiotensin I is further cleaved in the lungs by the ACE enzyme to form angiotensin II. Angiotensin II is one of the most potent vasoactive substances known to medicine, which means that it constricts blood vessels. When angiotensin II enters the bloodstream, the blood vessels get smaller and the heart tries to overcome this increased resistance by working more vigorously, causing the blood pressure to rise. Angiotensin II also causes the adrenal glands to releases aldosterone, a hormone that increases blood volume. The increased blood volume also acts to increase blood pressure.

Since the formation of blood-pressure-increasing angiotensin II is regulated by ACE, any compound that inhibits ACE will also block the formation of angiotensin II and might serve as a treatment for high blood pressure.

189: It was so profitable, in fact, that it made more money for Squibb than the rest of its drug portfolio combined: By 1985, there were multiple classes of antihypertensive therapies. The three most popular were the thiazide diuretics, the beta blockers, and the ACE inhibitors. The seemingly bottomless appetite for antihypertensive drugs led Pfizer to pursue a new class of antihypertensive. In 1985, scientists at the Pfizer laboratories in Sandwich, England, began studying a signaling molecule called cyclic GMP (cGMP) that was known to participate in multiple physiological pathways that controlled blood pressure. Even better, there appeared to be a well-established strategy for increasing cGMP levels: inhibiting the enzymes that degrade cGMP, enzymes called phosphodiesterases.

As Pfizer began searching for phosphodiesterase inhibitors, a trio of scientists announced their groundbreaking discovery that nitric oxide, a type of gas, served as a signaling molecule in the body. (This discovery would earn the scientists the 1998 Nobel Prize in medicine.) This finding had special implications for the treatment of angina pectoris, a form of chest pain produced by insufficient oxygen to the heart muscle. Ever since the nineteenth century, nitroglycerin was widely used to treat angina, but nobody knew how it worked. But now scientists realized that nitroglycerin released nitric oxide, which in turn caused blood vessels to dilate and supply more oxygen to the heart. Why did this matter to the Pfizer drug hunters? Because the second messenger for the action of nitric oxide turned out to be cGMP.

The Pfizer team in Sandwich switched their objective. They continued searching for a phosphodiesterase inhibitor that increased cGMP, but the goal was to develop a drug to treat angina pectoris instead of high blood pressure. In 1989, they finally found the right molecule: UK-92-480, which they later named sildenafil. In 1991, sildenafil entered clinical trials for angina . . . and the drug turned out to be a complete bust. It was not meaningfully better than nitroglycerin, the angina drug that had been discovered more than a century earlier and was available widely and cheaply.

However, a few of the scientists became intrigued by one of the side effects reported by the subjects of the clinical trials. Many of the male patients got erections.

At the time, there was very little in the way of treatment for erectile dysfunction. (In fact, the very phrase "erectile dysfunction" was not in widespread use.) Some physicians recommended a collection of pumps and constriction devices to aid with erections, although these were obviously not highly conducive to romance. There was a single approved drug for erectile dysfunction, alprostadil, but it had to be injected directly into the penis with a syringe or—even worse—pellets of the drug had to be shoved down the urethra of the penis. There were also prosthetic devices that needed to be surgically implanted. Thus, Pfizer concluded, there might be a very large market for a simple pill that helped men get erections.

Pfizer set up clinical trials for sildenafil as a treatment for erections. Almost nine out of ten of the male subjects (87.7 percent) stated that sildenafil improved their erections. An even larger majority wanted to continue using the drug. But perhaps most revealing was the feedback from the subjects. One wrote: "Before I took part in the study I was heavily depressed. I was continually arguing with my wife and generally making life hell for her and my children. . . . [E]ntering the study saved our family from much grief. . . . [I]t probably saved my marriage and possibly my life."

Another participant stated: "[T]he drug has proved very effective in enabling me to engage in sexual activity. . . despite my age (91) I am able to function as well as men many years my junior."

Pfizer filed the sildenafil application with the FDA in September 1997. The product was assigned for priority review and approval was announced on March 27, 1998. In 1998, Pfizer began selling its new drug, which it dubbed Viagra. Between 1998 and 2008, Pfizer reported $26 billion in total global sales of Viagra.

Chapter 11: The Pill: Drug Hunters Striking Gold outside of Big Pharma

216: Rock gathered together fifty female volunteers from his fertility laboratory and began giving them the three versions of the progesterone: The three versions were norethisterone (from Russell Marker's Syntex) and noretynodrel and norethandrolone (both made by Searle). In December of 1954, Rock began the first studies of the ovulation-suppressing potential of five 50 mg doses of the three oral progestins for three months (for twenty-one days per cycle; he administered the medication on days five through twenty-five followed by pill-free days in order to produce withdrawal bleeding). At 5 mg doses, norethisterone and noretynodrel—and norethandrolone at all doses—successfully suppressed ovulation, but they also caused breakthrough bleeding. At 10 mg or higher doses, norethisterone and noretynodrel suppressed ovulation without breakthrough bleeding and led to a 14 percent pregnancy rate during the following five months. Pincus and Rock selected Searle's noretynodrel for the contraceptive trials in Puerto Rico.

Chapter 12: Mystery Cures: Discovering Drugs through Blind Luck

232: most members of the psychiatric establishment believed there could *never* be a drug to treat these disorders: Woody Allen includes a classic joke in his film *Annie Hall*:

> It reminds me of that old joke—you know, a guy walks into a psychiatrist's office and says, hey doc, my brother's crazy! He thinks he's a chicken. Then the doc says, why don't you turn him in? Then the guy says, I would but I need the eggs. I guess that's how I feel about relationships. They're totally crazy, irrational, and absurd, but we keep going through it because we need the eggs.

Modern psychiatry has put extensive effort into distinguishing between mental illness and poor judgment. It's not easy, but there are two words for crazy used by German Jews that nicely capture the distinction: *meschugge* and *verrückt*. A middle-aged man, happily married for more than thirty years, suddenly becomes infatuated with his twenty-something secretary and starts an affair. His wife learns of it and demands a divorce. The man is immediately full of regret and apologizes, but the wife, feeling betrayed, refuses to reconcile. This is *meschugge*. In another case, a man hears his middle-aged brother has not been showing up for work and his neighbors report they have not seen him in days. The man enters his brother's home to find out what's going on and discovers him hiding under his bed, screaming and eating bugs. This is *verrückt*.

From a pharmacology perspective, most—if not all—psychiatric medications in current use are pretty poor drugs. In general, to proceed with the discovery of a new drug one needs either a validated pharmacological target or, lacking that, an animal model of the disease that can be used to test candidate drug compounds. The big problem with psychiatric disorders is that we still know very little about their physiological basis and can only guess at the neurochemical imbalances that underlie these diseases. Complicating matters even further is the fact that we cannot reproduce mental disorders in laboratory animals. How do we know for sure if an animal is feeling suicidal, suffering from hallucinations, or having disturbing thoughts? And how would we know if a drug is mitigating these abnormal thoughts and feelings?

236: The success of chlorpromazine also marked the beginning of the end for psychoanalysis: The widespread adoption of the first antipsychotic drugs soon led to the closing of mental hospitals around the country, a public health phenomenon known as deinstitutionalization. It was no longer necessary to warehouse the incurable mentally ill, since antipsychotic drugs enabled them to live within the community. Nevertheless, these drugs are highly imperfect medicines. Deinstitutionalization had the undesired and unintended effect of expelling many individuals whose psychotic disorders were only partially treated by drugs, or, even in cases where patients responded very well to antipsychotic drugs, many of these patients would stop taking their medication, especially since these drugs have many unpleasant side effects. Consequently, many deinstitutionalized patients have ended up in prison, which has now become the institution holding the highest number of mentally ill individuals. A 2011 *New England Journal of Medicine* article reports that the prevalence of mental health disorders among jail inmates is thirty times higher than in the general population. Jailing sick people is not an acceptable solution, and hopefully the discovery of new and more effective medicines will eventually settle this disturbing medical problem.

Conclusion: The Future of the Drug Hunter: The Chevy Volt and the Lone Ranger

250: More severely, in 2016, a painkiller drug trial in France killed one man and critically injured five others: Another tragic case: in 2006, TeGenero Immuno Therapeutics started clinical trials in London on TGN1412, a new drug that had been developed to treat leukemia and rheumatoid arthritis. It worked by modulating the human immune system. The drug was given to six healthy male volunteers at a tiny fraction of the dose that had previously been demonstrated to be safe in monkeys (one five-hundredth, or 0.2 percent). Within four hours all six men were gravely ill. They suffered catastrophic organ failure resulting from a "cytokine storm," which produces a massive flood of active immune cells and fluids. Four of the volunteers ended up in critical condition and almost died. Although all six volunteers eventually recovered, they may face a variety of immunological diseases later in life.

The British equivalent of the FDA, the Healthcare Products Regulatory Agency or MHRA, investigated the incident and found no evidence of fraud or malfeasance. TeGenero appeared to have honestly disclosed all their data to the regulatory authorities and followed the appropriate testing protocols. In response to the disaster there was a reevaluation of MHRA protocols. Consequently, the regulations for conducting clinical trials in Britain have been stiffened.

TeGenero filed for bankruptcy in late 2006.

Bibliography and Suggested Readings

Introduction: The Pharmaceutical Library of Babel

Ötzi the ice man

Fowler, Brenda. *Iceman: Uncovering the Life and Times of a Prehistoric Man Found in an Alpine Glacier*. Chicago: University of Chicago Press, 2001.

Rapamycin—Suren Sehgal

Loria, Kevin. "A Rogue Doctor Saved a Potential Miracle Drug by Storing Samples in His Home after Being Told to Throw Them Away." *Business Insider*, February 20, 2015.

Sehgal, S. N. "Sirolimus: Its Discovery, Biological Properties, and Mechanism of Action." *Transplant Procedures*. 35 (3 Suppl.) (2003): 7S–14S.

Seto, B. "Rapamycin and mTOR: A Serendipitous Discovery and Implications for Breast Cancer." *Clinical and Translational Medicine* 1 (2012): 1–29.

Cost of FDA-approved drug

DiMasi, J. A., H. G. Grabowski, and R. W. Hansen. "Innovation in the Pharmaceutical Industry: New Estimates of R&D Costs." Boston: Tufts Center for the Study of Drug Development, November 18, 2014. http://csdd.tufts.edu/news/complete_story/cost_study_press_event_webcast, retrieved January 4, 2016.

Emanuel, Ezekiel J. "Spending More Doesn't Make Us Healthier." *New York Times*, October 27, 2011.

"Research and Development in the Pharmaceutical Industry, A CBO Study." October 2006, https://www.cbo.gov/sites/default/files/109th-congress-2005-2006/reports/10-02-drugr-d.pdf, retrieved January 27, 2016.

Vagelos, P. R. "Are Prescription Prices Too High?," *Science* 252 (1991): 1080–4.

Possible shapes and charge distributions for compounds circa 500 daltons = 3 x 10^{62}

Bohacek, R. S., et al. "The Art and Practice of Structure-based Drug Design: A Molecular Modeling Perspective." Med. Res. Rev. 16 (1996): 3–50.

Library of Babel

Borges, Jorge Luis. *The Library of Babel.* Boston: David R. Godine, 2000.

Lipitor acts upon HMG-CoA reductase, the protein controlling the rate of cholesterol synthesis; penicillin shuts down peptidoglycan transpeptidase

Bruton, L., et al. Chapter 31, "Drug Therapy for Hypercholesterolemia and Dyslipidemia." In *Goodman and Gilman's The Pharmacological Basis of Therapeutics.* New York: McGraw-Hill Education/Medical (12th edition), 2011.

———. Chapter 53, "Penicillins, Cephalosporins, and Other β-Lactam Antibiotics." In *Goodman and Gilman's The Pharmacological Basis of Therapeutics.* New York: McGraw-Hill Education/Medical (12th edition), 2011.

Chloroform discovery

Dunn, P. M. "Sir James Young Simpson (1811–1870) and Obstetric Anesthesia." *Archives of Disease in Childhood, Fetal and Neonatal Edition* 86 (2002): F207–9.

Gordon, H. Laing. *Sir James Young Simpson and Chloroform (1811–1870).* New York: Minerva Group, 2002.

Drug discovery

Ravina, Enrique. *The Evolution of Drug Discovery.* Weinheim, Germany: Wiley-Verlag Helvetica Chimica, 2011.

Sneader, Walter. *Drug Discovery: A History*. Hoboken, NJ: John Wiley and Sons, 2005.

Chapter 1: So Easy a Caveman Can Do It: The Unlikely Origins of Drug Hunting

Opium

Booth, Martin. *Opium: A History*. London: St. Martin's Griffin, 2013.

Brownstein, M. J. "A Brief History of Opiates, Opioid Peptides, and Opioid Receptors," *Proceedings of the National Academy of Science USA* 90 (1993): 5391–3.

Goldberg, Jeff. *Flowers in the Blood: The Story of Opium*. New York: Skyhorse Publishing, 2014.

Hodgson, Barbara. *Opium: A Portrait of the Heavenly Demon*. Vancouver: Greystone Books, 2004.

Paracelsus and opium (laudanum)

Hodgson, Barbara. *In the Arms of Morpheus: The Tragic History of Morphine, Laudanum and Patent Medicines*. Richmond Hill: Firefly Books, 2001.

Paregoric

Boyd, E. M., and M. L. MacLachan. "The Expectorant Action of Paregoric." *Canadian Medical Association Journal* 50 (1944): 338–44.

Dover's Powder and Alexander Selkirk

Alleyel, Richard. "Mystery of Alexander Selkirk, the Real Robinson Crusoe, Solved." *Daily Telegraph*, October 30, 2008.

Bruce, J. S., and M. S. Bruce. "Alexander Selkirk: The Real Robinson Crusoe." *The Explorers Journal*, Spring 1993.

"Dr. Thomas Dover, Therapeutist and Buccaneer." *Journal of the American Medical Association*, February 29, 1896, 435.

Kraske, Robert, and Andrew Parker. *Marooned: The Strange but True Adventures of Alexander Selkirk, the Real Robinson Crusoe*. Boston: Clarion Books 2005.

Leslie, Edward E. "On a Piece of Stone: Alexander Selkirk on Greater Land." In *Desperate Journeys, Abandoned Souls: True Stories of Castaways and Other Survivors*. New York: Mariner Books 1998.

Osler, W. "Thomas Dover, M. B. (of Dover's Powder), Physician and Buccaneer." *Academy of Medicine* 82 (2007): 880–1.

Phear, D. N. "Thomas Dover 1662–1742; Physician, Privateering Captain, and Inventor of Dover's Powder." *Journal of the History of Medicine and Allied Sciences* 2 (1954) 139–56.

Selcraig, B. "The Real Robinson Crusoe." *Smithsonian Magazine*, July 2005.

Heroin and Bayer

Bruton et al. Chapter 18, "Opioids, Analgesia, and Pain Management." In *Goodman and Gilman's The Pharmacological Basis of Therapeutics*, New York: McGraw-Hill Education/Medical (12th edition), 2011.

Chemical Heritage Foundation Felix Hoffmann Biography, http://www.chemheritage.org/discover/online-resources/chemistry-in-history/themes/pharmaceuticals/relieving-symptoms/hoffmann.aspx, retrieved December 22, 2015.

Edwards, Jim. "Yes, Bayer Promoted Heroin for Children—Here Are the Ads That Prove It." *Business Insider*, November 17, 2011.

Scott, I. "Heroin: A Hundred Year Habit." *History Today*, vol. 48, 1998. http://www.historytoday.com/ian-scott/heroin-hundred-year-habit, retrieved January 27, 2016.

Sneader, W. "The Discovery of Heroin." *Lancet*, 352 (1998): 1697–9.

Sears Roebuck catalog heroin

Buxton, Julia. *The Political Economy of Narcotics*. London: Zed Books, 2013.

Endorphin receptor story

Terenius, L. "Endogenous Peptides and Analgesia." *Annual Review of Pharmacology and Toxicology* 18 (1978): 189–204.

Increase in THC in marijuana

Hellerman, C., "Is Super Weed, Super Bad?" CNN. http://www.cnn.com/2013/08/09/health/weed-potency-levels/, retrieved December 23, 2015.

Walton, A.G. "New Study Shows How Marijuana's Potency Has Changed Over Time." *Forbes*, March 23, 2015. http://www.forbes.com/sites/

alicegwalton/2015/03/23/pot-evolution-how-the-makeup-of-marijuana-has-changed-over-time/, retrieved December 23, 2015.

SCN9A (Nav1.7)

Drews, J., et al. "Drug Discovery: A Historical Perspective." *Science* 287 (2000): 1960-4.

King, G. F., and L. Vetter. "No Gain, No Pain: NaV1.7 as an Analgesic Target," *ACS Chemical Neuroscience* 5 (2014): 749–51.

Pina, A. S., et al. "An Historical Overview of Drug Discovery Methods." *Molecular Biology* 572 (2009): 3–12.

Chapter 2: Countess Chinchón's Cure: The Library of Botanical Medicine

Valerius Cordus, biography and discovery of ether

Arbor, Agnes. *"Herbals, Their Origin and Evolution: A Chapter in the History of Botany, 1470–1670."* Seattle: Amazon Digital Services, Inc., 1912.

Leaky, C. D. "Valerius Cordus and the Discovery of Ether." *Isis* 7 (1926): 14–24.

Sprague, T. A., and M. S. Sprague. "The Herbal of Valerius Cordus." *Journal of the Linnean Society of London.*52 (1939): 1–113.

Chinchona—history

Bruce-Chwatt, L. J. "Three Hundred and Fifty Years of the Peruvian Fever Bark." *British Medical Journal (Clinical Research Edition)* 296 (1988): 1486–7.

Butler A. R., et al. "A Brief History of Malaria Chemotherapy." *J R College of Physicians Edinborough* 40 (2010): 172–7.

Guerra, F. "The Introduction of Cinchona in the Treatment of Malaria." *Journal of Tropical Medicine and Hygiene* 80 (1977):112–18.

Humphrey, Loren. *Quinine and Quarantine: Missouri Medicine through the Years.* Missouri Heritage Readers. Columbia University of Missouri, 2000.

Kaufman T., and E. Rúveda. "The Quest for Quinine: Those Who Won the Battles and Those Who Won the War." *Angew Chemistry International Edition England* 44 (2005): 854–85.

Rocco, Fiammetta. *The Miraculous Fever-Tree: Malaria, Medicine and the Cure That Changed the World.* New York: HarperCollins, 2012.

———. *Quinine: Malaria and the Quest for a Cure That Changed the World*. New York: Harper Perennial, 2004.

Robert Talbor, quinine charlatan

"Jesuit's Bark" Catholic Encyclopedia 1913 https://en.wikisource.org/wiki/Catholic_Encyclopedia_(1913)/Jesuit%27s_Bark, retrieved December 29, 2015.

Keeble, T. A. "A Cure for the Ague: The Contribution of Robert Talbor (1642–81)," *Journal of the Royal Society of Medicine* 90 (1997): 285–90.

"Malaria." Royal Pharmaceutical Society, https://www.rpharms.com/museum-pdfs/c-malaria.pdf, retrieved December 24, 2015.

Talbor, Robert. *Pyretologia, A Rational Account of the Cause and Cure of Agues*. 1672.

Chapter 3: Standard Oil and Standard Ether: The Library of Industrial Medicine

George Wilson—foot amputation

"The Horrors of Pre-Anaesthetic Surgery." *Chirurgeon's Apprentice*, July 16, 2014. http://thechirurgeonsapprentice.com/2014/07/16/the-horrors-of-pre-anaesthetic-surgery/, retrieved December 29, 2015.

Lang, Joshua. "Awakening." *Atlantic*, January 2013. http://www.theatlantic.com/magazine/archive/2013/01/awakening/309188/, retrieved December 29, 2015.

Robert Liston, speed demon surgeon

Coltart, D. J. "Surgery between Hunter and Lister as Exemplified by the Life and Works of Robert Liston (1794–1847)." *Proceedings of the Royal Society of Medicine* 65 (1972): 556–60.

"Death of Robert Liston, ESQ., F.R.S.." *Lancet* 50 (1847): 633–4.

Ellis, Harold. *Operations That Made History*. Cambridge: Cambridge University Press, 2009.

Gordon, Richard. *Dr. Gordon's Casebook*. Cornwall: House of Stratus, 2001.

———. *Great Medical Disasters*. Cornwall House of Stratus, 2013.

Magee, R. "Surgery in the Pre-Anaesthetic Era: The Life and Work of Robert Liston." *Health and History* 2 (2000): 121–133.

William T. G. Morton and ether

Fenster, J. M. *Ether Day: The Strange Tale of America's Greatest Medical Discovery and the Haunted Men Who Made It.* New York: Harper Perennial, 2002.

"William T. G. Morton (1819–1868) Demonstrator of Ether Anesthesia." *JAMA.* 194 (1965): 170–1.

Wolfe, Richard, J. *Tarnished Idol: William Thomas Green Morton and the Introduction of Surgical Anesthesia.* Novato: Jeremy Norman Co; Norman Science-Technology, 2001.

John Collins Warren (Harvard Medical School) biography

Toledo, A. H. "John Collins Warren: Master Educator and Pioneer Surgeon of Ether Fame." *Journal of Investigative Surgery* 19 (2006): 341–4.

Warren, J. "Remarks on Angina Pectoris." *New England Journal of Medicine* 1 (1812): 1–11.

E. R. Squibb biography

"E. R. Squibb, Medical Drug Maker during the Civil War." http://www.medicalantiques.com/civilwar/Articles/Squibb_E_R.htm, retrieved January 4, 2016.

Rhode, Michael. "E. R. Squibb, 1854." *Scientist*, February 1, 2008.

Worthen, Dennis B. "Edward Robinson Squibb (1819–1900): Advocate of Product Standards." *Journal of the American Pharmaceutical Association* 46 (2003): 754–8.

———. *Heroes of Pharmacy: Professional Leadership in Times of Change.* Washington: American Pharmacists Association, 2012.

Chapter 4: Indigo, Violet, and Crimson: The Library of Synthetic Medicine

History of the German dye industry

Aftalion, Fred. *History of the International Chemical Industry: From the "Early Days" to 2000.* Philadelphia: Chemical Heritage Foundation, 2005.

Chandler, Alfred D. Jr. *Shaping the Industrial Century: The Remarkable Story of the Evolution of the Modern Chemical and Pharmaceutical Industries.* Cambridge: Harvard University Press (Harvard Studies in Business History), 2004.

Bayer: Duisberg, Eichengrün, Dreser

Biography Carl Duisberg, Bayer, http://www.bayer.com/en/carl-duisberg.aspx, retrieved January 4, 2016.

Rinsema, T. J. "One Hundred Years of Aspirin." *Medical History* 43 (1999): 502–7.

Sneader W. "The Discovery of Aspirin: A Reappraisal." *British Medical Journal* 321 (2000): 1591–4.

Aspirin history

Bruton, L. et al. Chapter 34, "Anti-inflammatory, Antipyretic, and Analgesic Agents; Pharmacotherapy of Gout." In *Goodman and Gilman's The Pharmacological Basis of Therapeutics*, New York: McGraw-Hill Education/Medical (12th edition), 2011.

Goodman, L. S. and A. Gilman. "Appendix" In *The Pharmacological Basis of Therapeutics*. New York: Macmillan, 1941.

Mahdi, J. G., et al. "The Historical Analysis of Aspirin Discovery, Its Relation to the Willow Tree and Antiproliferative and Anticancer Potential." *Cell Proliferation* 39 (2006): 147–55.

Vane, J. R. "Adventures and Excursions in Bioassay: The Stepping Stones to Prostacyclin." *British Journal of Pharmacology* 79 (1983): 821–38.

———. "Inhibition of Prostaglandin Synthesis as a Mechanism of Action for Aspirin-Like Drugs." *Nature New Biology* 231 (1971): 232–5.

Chapter 5: The Magic Bullet: We Figure Out How Drugs Actually Work

Syphilis history, symptoms

Harper, K. N., et al. "The Origin and Antiquity of Syphilis Revisited: An Appraisal of Old World Pre-Columbian Evidence for Treponemal Infection." *American Journal of Physical Anthropology* 146, Supplement 53 (2011): 99–133.

Kasper, D. et al. Chapter 206, "Syphilis." In *Harrison's Principles of Internal Medicine*. New York: McGraw-Hill Education/Medical (19th edition), 2015.

Miasma theory

Semmelweis, Ignaz. *Die Ätiologie der Begriff und die Prophylaxis des Kindbettfiebers* (The Etiology, Concept, and Prophylaxis of Childbed Fever), 1861.

Louis Pasteur

Birch, Beverly, and Christian Birmingham. *Pasteur's Fight against Microbes (Science Stories)*. Hauppauge: Barron's Educational Series, 1996.

Tiner, John Hudson. *Louis Pasteur: Founder of Modern Medicine.* Fenton: Mott Media, 1999.

Paul Ehrlich biography and Salvarsan

Sepkowitz, K. A. "One Hundred Years of Salvarsan." *New England Journal of Medicine* 365 (2011): 291–3.

Receptor theory history and Ehrlich's reaction to counter theories

Prüll, Cay-Ruediger, et al. *A Short History of the Drug Receptor Concept (Science, Technology & Medicine in Modern History)*. Basingstoke: Palgrave Macmillan, 2009.

Chapter 6: Medicine That Kills: The Birth of Drug Regulation

Avorn, J. "Learning About the Safety of Drugs—A Half-Century of Evolution." *New England Journal of Medicine*, 365 (2011): 2151–3.

Bayer and the Prontosil story

Bentley, R. "Different Roads to Discovery; Prontosil (Hence Sulfa Drugs) and Penicillin (Hence Beta-Lactams)." *Journal Industrial Microbiology and Biotechnology* 36 (2009): 775–86.

Hager, Thomas. *The Demon under the Microscope: From Battlefield Hospitals to Nazi Labs, One Doctor's Heroic Search for the World's First Miracle Drug*. New York: Broadway Books, 2007.

Otten, H. "Domagk and the Development of the Sulphonamides." *Journal of Antimicrobial Chemotherapy* 17 (1986): 689–96.

Prodrug: sulfanilamide

Lesch, John E. *The First Miracle Drugs: How the Sulfa Drugs Transformed Medicine.* Oxford: Oxford University Press, 2006.

S. E. Massengill and elixir sulfanilamide

Akst, J. "The Elixir Tragedy, 1937." *Scientist*, June 1, 2013.

"Deaths Following Elixir of Sulfanilamide-Massengill" *Journal of the American Medical Association* 109 (1937): 1610–11.

Geiling, E. M. K., and P. R. Cannon. "Pathological Effects of Elixir of Sulfanilamide (Diethylene Glycol) Poisoning," *Journal of the American Medical Association* 111 (1938): 919–926.

Wax, P. M. "Elixirs, Diluents, and the Passage of the 1938 Federal Food, Drug and Cosmetic Act." *American College of Physicians* 122 (1995): 456–61.

FDA reaction for elixir sulfanilamide

Ballentine. C. "Sulfanilamide Disaster." *FDA Consumer Magazine,* June 1981, http://www.fda.gov/aboutfda/whatwedo/history/productregulation/sulfanilamidedisaster/default.htm, retrieved January 4, 2016.

"Elixir of Sulfanilamide: Deaths in Tennessee." *Pathophilia for the Love of Disease.* http://bmartinmd.com/eos-deaths-tennessee/, retrieved January 4, 2016.

ACT UP—AIDS

Crimp. D. "Before Occupy: How AIDS Activists Seized Control of the FDA in 1988," *Atlantic,* December 6, 2011.

Fen-Phen

Connolly, H. M., et al. "Valvuolar Heart Disease Associated with Fenfluramine–Phentermine." *New England Journal of Medicine* 337 (1997): 581–8.

Courtwright, D. T. "Preventing and Treating Narcotic Addiction—A Century of Federal Drug Control." *New England Journal of Medicine* 373: (2015) 2095–7.

Chapter 7: The Official Manual of Drug Hunting: Pharmacology Becomes a Science

Goodman biography and Gilman biography (research accomplishments—cancer, curare)

Altman, Lawrence K. "Dr. Louis S. Goodman, 94, Chemotherapy Pioneer, Dies." *New York Times,* November 28, 2000.

Ritchie, M. "Alfred Gilman: February 5, 1908–January 13, 1984." *Biographies of Members of the National Academy of Science* 70 (1996): 59–80.

Clark Stanley, snake oil king biography

Dobie, J. Frank. *Rattlesnakes*. Austin: University of Texas Press, 1982.

"A History of Snake Oil Salesmen." http://www.npr.org/sections/codeswitch/2013/08/26/215761377/a-history-of-snake-oil-salesmen, retrieved January 8, 2016.

"Why Are Snake-Oil Remedies So-Called?" http://www.canada.com/montrealgazette/news/books/story.html?id=666775cc-f9ff-4360-9533-4ea7f0eef233, retrieved January 8, 2016.

History of teaching pharmacology in medical schools (Abraham Flexner)

Bonner, Thomas Neville. *Iconoclast: Abraham Flexner and a Life in Learning*. Baltimore: Johns Hopkins University Press, 2002.

Chapter 8: Beyond Salvarsan: The Library of Dirty Medicine

Isak Dinesen biography

Dinesen, Isak. *Out of Africa: And Shadows on the Grass*. New York: Vintage Books, 2011.

Hannah, Donald. *Isak Dinesen and Karen Blixen: The Mask and the Reality*. New York: Random House, 1971.

Alexander Fleming, Abraham Chain, and Howard Florey biographies and papers

Abraham, Edward P. "Ernst Boris Chain. 19 June 1906–12 August 1979." *Biographical Memoirs of Fellows of the Royal Society* 29 (1983): 42–91.

———. "Howard Walter Florey. Baron Florey of Adelaide and Marston 1898–1968." *Biographical Memoirs of Fellows of the Royal Society* 17 (1971): 255–302.

Brown, Kevin. *Penicillin Man: Alexander Fleming and the Antibiotic Revolution*. Dublin: History Press Ireland, 2013.

Chain, E., et al. "Further Observations on Penicillin." *Lancet*, August 16, 1941, 177–88.

———. "Penicillin as a Chemotherapeutic Agent." *Lancet*, August 20, 1940 226–28.

Colebrook, L. "Alexander Fleming 1881–1955." *Biographical Memoirs of Fellows of the Royal Society* 2 (1956): 117–27.

Lax, Eric. *The Mold in Dr. Florey's Coat: The Story of the Penicillin Miracle*. New York: Henry Holt and Company, 2015.

Macfarlane, Gwyn. *Alexander Fleming: The Man and the Myth*. Cambridge: Harvard University Press, 1984.

———. *Howard Florey: The Making of a Great Scientist*. Oxford: Oxford University Press 1979.

Mazumdar, P. M. "Fleming as Bacteriologist: Alexander Fleming." *Science* 225 (1984): 1140.

Raju, T. N. "The Nobel Chronicles. 1945: Sir Alexander Fleming (1881–1955); Sir Ernst Boris Chain (1906–79); and Baron Howard Walter Florey (1898–1968)." *Lancet* 353 (1999): 936.

Shampo, M. A. and R. A. Kyle. "Ernst Chain—Nobel Prize for Work on Penicillin." *Mayo Clinic Proceedings* 75 (2000): 882.

"Sir Howard Florey, F.R.S.: Lister Medallist." *Nature* 155 (1945): 601.

History of penicillin

Bud, Robert. *Penicillin: Triumph and Tragedy*. Oxford: Oxford University Press, 2009.

Hare, R. "New Light on the History of Penicillin." *Medical History* 26 (1982): 1–24.

Selman Waksman biography—streptomycin

Hotchkiss, R. D. "Selman Abraham Waksman." *Biographies of Members of the National Academy of Science* 83 (2003): 320-43.

Pringle, Peter. *Experiment Eleven: Dark Secrets Behind the Discovery of a Wonder Drug*. London: Walker Books, 2012.

"Selman A. Waksman (1888–1973)." http://web.archive.org/web/20080418134324/http://waksman.rutgers.edu/Waks/Waksman/DrWaksman.html, retrieved January 6, 2016.

Wainwright, M. "Streptomycin: Discovery and Resultant Controversy." *History and Philosophy of the Life Sciences* 13: (1991) 97–124.

Waksman, Selman A. *My Life with the Microbes*, New York: Simon and Schuster, 1954.

History of TB

Bynum, Helen. *Spitting Blood: The History of Tuberculosis.* Oxford: Oxford University Press, 2015.

Dormandy, Thomas. *The White Death: A History of Tuberculosis.* New York: New York University Press, 2000.

Goetz, Thomas. *The Remedy: Robert Koch, Arthur Conan Doyle, and the Quest to Cure Tuberculosis.* New York: Gotham, 2014.

Golden age antibiotic discovery

Demain, A. L. "Industrial Microbiology." *Science* 214 (1981): 987–95.

Chapter 9: The Pig Elixir: The Library of Genetic Medicine

History of insulin

Baeshen, N.A., et al. "Cell Factories for Insulin Production." *Microbial Cell Factories* 13 (2014): 141–150.

Bliss, Michael. *Banting: A Biography*. Toronto: University of Toronto Press, Scholarly Publishing Division, 1993.

Bliss, Michael. *The Discovery of Insulin*. Chicago: University Of Chicago Press, 2007.

Cooper, Thea, and Arthur Ainsberg. *Breakthrough: Elizabeth Hughes, the Discovery of Insulin, and the Making of a Medical Miracle.* London: St. Martin's Griffin, 2011.

Mohammad K., M. K. Ghazavi, and G. A. Johnston. "Insulin Allergy." *Clinics in Dermatology* 29 (2011): 300–305.

History of Eli Lilly

Manufacturing Pharmaceuticals: Eli Lilly and Company, 1876–1948. In James Madison, *Business and Economic History*, 1989. Business History Conference.

History of diabetes

Auwerx, J. "PPARgamma, the Ultimate Thrifty Gene." *Diabetalogia* 42 (1999): 1033–49.

Blades M., et al. "Dietary Advice in the Management of Diabetes Mellitus—History and Current Practice." *Journal of the Royal Society of Health* 117 (1997): 143–50.

Brownson, R. C., et al. "Declining Rates of Physical Activity in the United States:

What Are the Contributors?" *Annual Review of Public Health* 26 (2005): 421–43.

Brunton, L,. et al. Chapter 43, "Endocrine Pancreas and Pharmacotherapy of Diabetes Mellitus and Hypoglycemia." In *Goodman and Gilman's The Pharmacological Basis of Therapeutics*. New York: McGraw-Hill Education/ Medical (12th edition), 2011.

Duhault, J., and R. Lavielle. "History and Evolution of the Concept of Oral Therapy in Diabetes." *Diabetes Research and Clinical Practice,* 14 suppl 2 (1991): S9–13.

Eknoyan, G., and J. Nagy. "A History of Diabetes Mellitus or How a Disease of the Kidneys Evolved into a Kidney Disease." *Advances in Chronic Kidney Disease* 12 (2005) : 223–9.

Ezzati, M., and E. Riboli. "Behavioral and Dietary Risk Factors for Noncom municable Diseases." *New England Journal of Medicine* 369 (2013): 954–64.

Gallwitz, B. "Therapies for the Treatment of Type 2 Diabetes Mellitus Based on Incretin Action." *Minerva Endocrinology* 31 (2006): 133–47.

Güthner, T., et al. "Guanidine and Derivatives." In *Ullmann's Encyclopedia of Industrial Chemistry*. Weinheim, Germany: Wiley-Verlag Helvetica Chimica, 2010.

Hoppin, A. G., et al. "Case 31-2006: A 15-Year-Old Girl with Severe Obesity." *New England Journal of Medicine* 355 (2006): 1593–1602.

Kasper, D., et al. Chapter 417, "Diabetes Mellitus: Diagnosis, Classification, and Pathophysiology." In *Harrison's Principles of Internal Medicine.*19th edition. New York: McGraw-Hill Education, 2015.

Kleinsorge, H. "Carbutamide—The First Oral Antidiabetic. A Retrospect." *Experimental Clinical Endocrinology and Diabetes* 106 (1998): 149–51.

Loubatieres-Mariani, M. M. "[The Discovery of Hypoglycemic Sulfonamides—original article in French]." *Journal of the Society of Biology* 201 (2007): 121–5.

Mogensen, C. E. "Diabetes Mellitus: A Look at the Past, a Glimpse to the Future." *Medicographia* 33 (2011): 9–14.

Parkes, D. G., et al. "Discovery and Development of Exenatide: the First Antidiabetic Agent to Leverage the Multiple Benefits of the Incretin Hormone, GLP-1." *Expert Opinion in Drug Discovery* 8 (2013): 219–44.

Pei, Z. "From the Bench to the Bedside: Dipeptidyl Peptidase IV Inhibitors, a

New Class of Oral Antihyperglycemic Agents." *Current Opinion in Discovery and Development* 11 (2008): 512–32.

Slotta, K. H., and T. Tschesche. "Uber Biguanide. II. Die Blutzucker senkende Wirkung der Biguanides." *Berichte der Deutschen Chemischen Gesellschaft B: Abhandlungen*, 62 (1929): 1398–1405.

Staels, B., et al. "The Effects of Fibrates and Thiazolidinediones on Plasma Triglyceride Metabolism Are Mediated by Distinct Peroxisome Proliferator Activated Receptors (PPARs)." *Biochemie* 79 (1997): 95–9.

Thornberry, N, A., and A. E. Weber. "Discovery of JANUVIA (Sitagliptin), a Selective Dipeptidyl Peptidase IV Inhibitor for the Treatment of Type 2 Diabetes." *Current Topics in Medicinal Chemistry* 7 (2007): 557–68.

Yki-Jarvinen, H. "Thiazolidinediones." *New England Journal of Medicine* 351 (2004): 1106–18.

History of insulin

Poretsky, Leonid. *Principles of Diabetes Mellitus*, New York: Springer, 2010.

Sönksen, P. H. "The Evolution of Insulin Treatment." *Clinical Endocrinology and Metabolism* 6 (1977): 481–97.

Lilly as insulin manufacturer

Madison, James, H. *Eli Lilly: A Life, 1885–1977*, Indianapolis: Indiana Historical Society, 2006.

History of gene cloning

Tooze, James, and John Watson. *The DNA Story: A Documentary History of Gene Cloning*. New York: W. H. Freeman, 1983.

Development of the biotech industry

Hughes, Sally Smith. *Genentech: The Beginnings of Biotech*. Chicago: University of Chicago Press, 2011.

Leaser, B., et al. "Protein Therapeutics: A Summary and Pharmacological Classification," *Nature Review Drug Discovery* 7 (2008): 21–39.

Shimasaki, Craig, ed. *Biotechnology Entrepreneurship: Starting, Managing, and Leading Biotech Companies*. San Diego: Academic Press, 2014.

Recombinant monoclonal antibodies

Marks, Lara V. *The Lock and Key of Medicine: Monoclonal Antibodies and the Transformation of Healthcare*. New Haven: Yale University Press, 2015.

Shire, Stephen. *Monoclonal Antibodies: Meeting the Challenges in Manufacturing, Formulation, Delivery and Stability of Final Drug Product*. Sawston, Cambridge: Woodhead Publishing, 2015

Chapter 10: From Blue Death to Beta Blockers: The Library of Epidemiological Medicine

John Snow biography

Hempel, Sandra. *The Strange Case of the Broad Street Pump: John Snow and the Mystery of Cholera*. Oakland: University of California Press, 2007.

Johnson, Steven. *The Ghost Map: The Story of London's Most Terrifying Epidemic—and How It Changed Science, Cities, and the Modern World*. New York: Riverhead Books, 2006.

Cholera background and history

Gordis, Leon. *Epidemiology*, Philadelphia, PA: Saunders, 2008.

Kotar, S. L. and G. E. Gessler. *Cholera: A Worldwide History*. Jefferson: McFarland & Company, 2014.

Polio and sugar story

Nathanson, N. and O. M. Kew. "From Emergence to Eradication: The Epidemiology of Poliomyelitis Deconstructed." *American Journal of Epidemiology* 172 (2010): 1213–29.

Framingham heart study

Bruenn, H. G. "Clinical Notes on the Illness and Death of President Franklin D. Roosevelt." *Annals Internal Medicine* 72 (1970): 579–91.

Hay, J. H. "A British Medical Association Lecture on THE SIGNIFICANCE OF A RAISED BLOOD PRESSURE." *British Medical Journal* 2: (1931) 43–47.

Kolata, G. "Seeking Clues to Heart Disease in DNA of an Unlucky Family." *New York Times*, May 12, 2013.

Levy, Daniel. "60 Years Studying Heart-Disease Risk." *Nature Reviews Drug Discovery* 7 (2008): 715.

———. *Change of Heart: Unraveling the Mysteries of Cardiovascular Disease*. New York: Vintage Books, 2007.

Mahmood, S. S., et al. "The Framingham Heart Study and the Epidemiology of Cardiovascular Disease: A Historical Perspective." *Lancet* 383 (2014): 999–1008.

Hypertension history

Esunge, P. M. "From Blood Pressure to Hypertension: The History of Research." *Journal of the Royal Society of Medicine* 84 (1991): 621.

Postel-Vinay, Nicolas, ed., *A Century of Arterial Hypertension: 1896–1996*, Hoboken: Wiley, 1997.

History of hydrochlorothiazide

Beyer, K. H. "Chlorothiazide: How the Thiazides Evolved as Anti-Hypertensive Therapy." *Hypertension* 22 (1993): 388–91.

Burkhart, Ford. "Dr. Karl Beyer Jr., 82, Pharmacology Researcher." *New York Times*, December 16, 1996.

James Black biography and beta blocker

Black J. W. et al. "A New Adrenergic Beta Receptor Antagonist." *Lancet* 283 (1964): 1080–1.

Scheindlin, S. "A Century of Ulcer Medications," *Molecular Interventions* 5 (2005): 201–6

Sir James W. Black, Biographical, http://www.nobelprize.org/nobel_prizes/medicine/laureates/1988/black-bio.html, retrieved January 9, 2016.

Cushman and Ondetti

Cushman, D. W., and M. A. Ondetti. "History of the Design of Captopril and Related Inhibitors of Angiotensin Converting Enzyme," *Hypertension* 17 (1991): 589–92.

Ondetti, Miguel. https://en.wikipedia.org/wiki/Miguel_Ondetti, retrieved January 4, 2016.

Ondetti, Miguel, et al. "Design of Specific Inhibitors of Angiotensin-Converting Enzyme: New Class of Orally Active Anti-Hypertensive Agents." *Science*, new series 196 (1977): 441–4.

Smith, C. G., and J. R. Vane. "The Discovery of Captopril." FASEB Journal 17 (2003): 788–9.

Cholesterol and heart disease

Alberts, A. W. "Discovery, Biochemistry and Biology of Lovastatin." *American Journal of Cardiology* 62 (1988): 10J–15J.

Kolata, G. "Cholesterol-Heart Disease Link Illuminated," *Science* 221 (1983): 1164–6.

Tobert, J. A. "Lovastatin and Beyond: The History of the HMG-CoA Reductase Inhibitors." *Nature Reviews Drug Discovery* 2 (2003): 517–26.

Vaughn, C. J., et al. "The Evolving Role of Statins in the Management of Atherosclerosis." *Journal of the American College Cardiology* 35 (2000): 1–10.

Joseph Goldstein and Michael Brown familial hypercholesterolemia

Brown, M. S., and J. L. Goldstein. "A Receptor Mediated Pathway for Cholesterol Homeostasis." http://www.nobelprize.org/nobel_prizes/medicine/laureates/1985/brown-goldstein-lecture.pdf, retrieved January 9, 2016.

History of statins

Smith G. D., and J. Pekkanen. "The Cholesterol Controversy." *British Medical Journal* 304 (1992): 913.

Chapter 11: The Pill: Drug Hunters Striking Gold Outside of Big Pharma

History of the pill: hormones and ovulation history

Eig, Jonathan. *The Birth of the Pill: How Four Crusaders Reinvented Sex and Launched a Revolution.* New York: W. W. Norton, 2015.

Goldzieher, J. W., and H. W. Rudel. "How the Oral Contraceptives Came to be Developed." *Journal of the American Medical Association* 230 (1974): 421–5.

Russell Marker biography

Lehmann, P. A., et al. "Russell E. Marker Pioneer of the Mexican Steroid Industry." *Journal of Chemical Education* 50 (1973): 195–9.

Marker degradation

"The 'Marker' Degradation and the Creation of the Mexican Steroid Hormone Industry 1938–1945." American Chemical Society. https://www.acs.org/content/dam/acsorg/education/whatischemistry/landmarks/progesteronesynthesis/marker-degradation-creation-of-the-mexican-steroid-industry-by-russell-marker-commemorative-booklet.pdf, retrieved January 4, 2016.

Syntex

Laveaga, Gabriela Soto. *Jungle Laboratories: Mexican Peasants, National Projects, and the Making of the Pill.* Durham: Duke University Press, 2009.

Gregory Pincus

Diczfalusy, E. "Gregory Pincus and Steroidal Contraception: A New Departure in the History of Mankind." *Journal of Steroid Biochemistry* 11 (1979): 3–11.

"Dr. Pincus, Developer of Birth-Control Pill, Dies." *New York Times,* August 23, 1967.

Baron de Hirsch Fund

Joseph, Samuel. *History of the Baron De Hirsch Fund: Americanization of the Jewish Immigrant.* Philadelphia: Jewish Publication Society, 1935; New York: Augustus M. Kelley Publishing, January 1978.

Margaret Sanger

Chesler, Ellen. *Woman of Valor: Margaret Sanger and the Birth Control Movement in America.* New York: Simon & Schuster, 2007.

Grant, George, and Kent Hovind. *Killer Angel: A Short Biography of Planned Parenthood's Founder, Margaret Sanger.* Amazon Digital Services, 2015.

Sanger, Margaret. *The Autobiography of Margaret Sanger,* Mineola: Dover Publications, 2012.

Katharine Dexter McCormick

Engel, Keri Lynn. "Katharine McCormick, Biologist and Millionaire Philanthropist." Amazing Women in History http://www.amazingwomeninhistory.com/katharine-mccormick-birth-control-history/, retrieved January 3, 2016.

John Rock

Berger, Joseph. "John Rock, Developer of the Pill and Authority on Fertility, Dies." *New York Times*, December 5, 1984.

Gladwell, Malcolm. "John Rock's Error." *The New Yorker*, March 13, 2000.

Chapter 12: Mystery Cures: Discovering Drugs through Blind Luck

James Lind and scurvy

Gordon, E. C. "Scurvy and Anson's Voyage Round the World: 1740–1744. An Analysis of the Royal Navy's Worst Outbreak." *American Neptune* 44 (1984): 155–166.

Lamb, Jonathan. "Captain Cook and the Scourge of Scurvy." http://www.bbc.co.uk/history/british/empire_seapower/captaincook_scurvy_01.shtml, retrieved February 20, 2016.

McNeill, Robert B. *James Lind: The Scot Who Banished Scurvy and Daniel Defoe, England's Secret Agent*. Amazon Digital Services, 2011.

George Rieveschl biography

"The George Rieveschl, Jr., Papers (January 9, 1916–September 27, 2007), Collection No. 19." http://www.lloydlibrary.org/archives/inventories/rieveschl_papers_finding_aid.pdf, retrieved January 4, 2016.

Hevesli, D. "George Rieveschl, 91, Allergy Reliever, Dies." *New York Times*, September 29, 2007.

Muller, G. "Medicinal Chemistry of Target Family-Directed Masterkeys." *Drug Discovery Today*. 8 (2003): 681–91.

Diphenhydramine

Brunton, Laurence, et al., eds. Chapter 32, "Histamine, Bradykinin, and Their Antagonists." In *Goodman and Gilman's The Pharmacological Basis of Therapeutics*, New York: McGraw-Hill Education/Medical (12th edition), 2011.

Schizophrenia and Chlorpromazine

Ban, T. A. "Fifty Years Chlorpromazine: A Historical Perspective." *Neuropsychiatric Disease and Treatment* 3 (2007) : 495500.

Freedman, R. "Schizophrenia." *New England Journal of Medicine* 349 (2003): 1738–49.

Lieberman, Jeffrey A. *Shrinks: The Untold Story of Psychiatry*. New York: Little, Brown and Company, 2015.

Moussaoui, Driss. "A Biography of Jean Delay: First President of the World Psychiatric Association (History of the World Psychiatric Association)." *Excerpta Medica*, 2002.

Nasar, Sylvia. *A Beautiful Mind*. New York: Simon & Schuster, 2011.

"Paul Charpentier, Henri-Marie Laborit, Simone Courvoisier, Jean Delay, and Pierre Deniker." Chemical Heritage Foundation. http://www.chemheritage.org/discover/online-resources/chemistry-in-history/themes/pharmaceuticals/restoring-and-regulating-the-bodys-biochemistry/charpentier--laborit--courvoisier--delay--deniker.aspx, retrieved January 4, 2016.

Roland Kuhn and depression—Geigy relationship

Belmaker, R. H., and G. Agam. "Major Depressive Disorder." *New England Journal of Medicine* (358, 2008): 55–68.

Bossong, F. "Erinnerung an Roland Kuhn (1912–2005) und 50 Jahre Imipramin." *Der Nervenarzt* 9 (2008): 1080.

Cahn, Charles. "Roland Kuhn, 1912–2005." *Neuropsychopharmacology* 31 (2006): 1096.

Imipramine

Ayd, Frank J., and Barry Blackwell. Ayd. *Discoveries in Biological Psychiatry*. Philadelphia: J. B. Lippincott, 1970.

Fangmann, P., et al. "Half a Century of Antidepressant Drugs." *Journal of Clinical Psychopharmacology* 28 (2008): 1–4.

Shorter, Edward. *Before Prozac: The Troubled History of Mood Disorders in Psychiatry*. Oxford: Oxford University Press, 2008.

———. *A Historical Dictionary of Psychiatry*. Oxford: Oxford University Press, 2005.

Conclusion: The Future of the Drug Hunter: The Chevy Volt and the Lone Ranger

Chevy Volt history

Edsall, Larry. *Chevrolet Volt: Charging into the Future*. Minneapolis: Motorbooks, 2010.

Viagra—sildenafil

Ghofrani, H. A., et al. "Sildenafil: From Angina to Erectile Dysfunction to Pulmonary Hypertension and Beyond." *Nature Review Drug Discovery* 5 (2006): 689–702.

Cialis—tadalafil

Rotella, D. P. "Phosphodiesteras 5 Inhibitors: Current Status and Potential Applications." *Nature Review Drug Discovery* 1 (2002): 674–82.

NovoBiotic Pharmaceuticals story

Grady, Dennis. "New Antibiotic Stirs Hope Against Resistant Bacteria." *New York Times*, January 7, 2015.

Kaeberlin, T., et al. "Isolating 'Uncultivable' Microorganisms in Pure Culture in a Simulated Natural Environment." *Science* 296 (2002): 1127–9.

Naik, Gautam. "Scientists Discover Potent Antibiotic, A Potential Weapon Against a Range of Diseases." *Wall Street Journal*, January 9, 2015.

Index